THOMAS M. DISCH

THE DREAMS OUR STUFF IS MADE OF

How Science Fiction Conquered the World

A TOUCHSTONE BOOK
Published by Simon & Schuster
New York London Toronto Sydney Singapore

TOUCHSTONE
Rockefeller Center
1230 Avenue of the Americas
New York, NY 10020

First Touchstone Edition 2000

TOUCHSTONE and colophon are registered trademarks
of Simon & Schuster, Inc.

Designed by Pei Koay

Manufactured in the United States of America

10 9 8 7 6 5 4 3 2 1

The Library of Congress has cataloged the Free Press edition as follows:

Disch, Thomas M.
The dreams our stuff is made of : how science fiction conquered the world / Thomas M. Disch.
p. cm.
Includes bibliographical references and index.
1. Science fiction—History and criticism. 2. Literature and society. I. Title.
PN3433.5.D57 1998
809.3'8762—dc21 97-52068
CIP

ISBN 0-684-82405-1
0-684-85978-5 (Pbk)

Additional praise for *The Dreams Our Stuff Is Made Of*

"[Disch speaks] for the entire genre with admirable authority and grace."
—Alyssa Katz, *Newsday*

"Provocative."
—Christopher Lehmann-Haupt, *The New York Times*

"Entertaining and provocative. . . . Long overdue—a pioneering look at what other cultural observers have ignored. . . . Disch convinces us that any cultural study of our century which does not deal seriously with the genre is or will be fatally flawed."
—Dick Allen, *The Hudson Review*

"A provocative and enjoyable book."
—Robert Taylor, *The Boston Globe*

"Sharp, provocative. . . . More than just a history, Disch gives us a sense of the events and moods that are so much a part of science-fiction. . . . [Disch] has covered the vital aspects of the field in a highly readable book."
—Robert Sheckley, *San Francisco Chronicle*

"Fascinating."
—Michael Jacobs, *USA Today*

"A witty tour. . . . Disch is funny and provocative."
—Fred Cleaver, *The Denver Post*

"Absolutely fascinating. . . . Combines wit and scholarship with a pointed critical impulse. . . . Disch will force you to give up your prejudices."
—*Forbes*

"A rollicking thought-provoker of a book."
—Ray Olson, *Booklist*

"With pungency and wit, Disch explores the enormous cultural impact that SF [science fiction] has had over the past century."
—*Publishers Weekly*

"A rip-roaring ride. . . . Disch's account is sure to raise the hackles of many while providing an insightful study. Recommended."
—Ronald Ratliff, *Library Journal*

"A gifted writer casts a critical eye on the genre that gave him birth. . . . Disch's provocative, engrossing book may fan the flames of a number of simmering arguments in the SF community, but when the smoke clears we may all, as a result of this tonic work, see more clearly."
—*Kirkus Reviews*

"Thomas Disch, always ruefully wise and exuberantly visionary, gives us a superbly disturbing meditation on the triumph of science fiction over our expiring (or is it expired?) culture."

—Harold Bloom

"Thomas M. Disch's tough love survey of America's most characteristic pulp genre is one of those rare birds—an authentic critical work that's muscular, smart, chatty, and a pleasure to read. If you want to find out what's been happening to our weirdly imagined future for the past century or so, don't bother watching the skies—they're filled with the same big-screen nonsense and billboard bull as always. Just read this book and find out what's been really going on."

—Scott Bradfield, author of *The History of Luminous Motion*

"Disch is one of the Secret Masters of science fiction: knowing, masterful, sly, hilarious, and profound. This bible of SF insight, this devil's dictionary of sharp wisdom is the best book on SF ever written by a practicing writer in the field."

—John Clute, co-editor of *The Encyclopedia of Science Fiction*

"Disch has written an elegant and wonderfully lucid study of science fiction, in all its glorious permutations. Now that we have this definitive book, a subject too often skimmed or oversimplified has been given its due and then some."

—Nicholas Christopher, author of *Somewhere in the Night: Film Noir and the American City*

By the Same Author

NOVELS

The Genocides
The Puppies of Terra
Echo Round His Bones
Black Alice (with John Sladek)
Camp Concentration
334
Clara Reeve
On Wings of Song
Neighboring Lives (with Charles Naylor)
The Businessman: A Tale of Terror
The M.D.: A Horror Story
The Priest: A Gothic Romance

SHORT STORY COLLECTIONS

102 H-Bombs
Fun With Your New Head
Getting Into Death
Fundamental Disch
The Man Who Had No Idea

POETRY COLLECTIONS

The Right Way To Figure Plumbing
ABCDEFGHIJKLMNPOQRSTUVWXYZ
Orders of the Retina
Burn This
Here I Am, There You Are, Where Were We
Yes, Let's: New and Selected Poems
Dark Verses and Light

CHILDREN'S BOOKS

The Brave Little Toaster
The Tale of Dan de Lion
The Brave Little Toaster Goes to Mars
A Child's Garden of Grammar

LIBRETTI AND PLAYS

The Fall of the House of Usher (composer, Gregory Sandow)
Frankenstein (composer, Sandow)
Ben Hur
The Cardinal Detoxes

CRITICISM

The Castle of Indolence: On Poetry, Poets and Poetasters

INTERACTIVE SOFTWARE

Amnesia

Dedicated to the memory

of Glenn Wright

CONTENTS

INTRODUCTION

There used to be a truism—I heard it first from my then agent Terry Carr in 1964—that the golden age of science fiction is twelve, the age we begin to read SF and are wonderstruck. That truism is no longer true, for science fiction has come to permeate our culture to such a degree that its basic repertory of images—rocket ships and robots, aliens and dinosaurs—are standard items in the fantasy life of any preschooler. As for the twelve-year-olds of our own era, nothing science-fictional is alien to them.

Admittedly, theirs is not the science fiction of the printed page, for today's twelve-year-olds have been warped away from the Guttenberg galaxy. Instead of graduating from comic books to pulp magazines, as Terry and I did, the brighter children of the 1990s transfer their attention from the TV screen to the computer monitor as they mature. In both media, it is increasingly difficult to distinguish between science fiction and assorted neighboring realities. The dinosaurs in the movies look as real as elephants or camels; toddlers' toys morph into weapons; grown-ups on talk shows discuss their UFO abductions, while on the next channel a dull documentary recounts the history of space exploration. One has to be sent to school to begin to sort out what's real and what's Hollywood.

Science fiction is one of the few American industries that has never been transplanted abroad with any success. Japan may have zapped Detroit, but most sci fi still bears the label "Made in America," and the future represented by SF writers continues to be an American future. It isn't only Oz that is Kansas in disguise; the whole Galactic Imperium is simply the American Dream (or Nightmare) writ large. British SF writers decorated their stories with American slang just as their rock stars imitated American accents. When French film directors like Truffaut and Besson make SF movies, they set them in American cities.

The American SF dream first began to coalesce, as an institution and an industry, some seventy years ago, with the appearance of *Amazing Stories*, the first pulp magazine specializing in "scientifiction," as Hugo Gernsback christened the then nameless genre. From its inception, American SF has been a naive, ungainly hybrid, full of inconsistencies and obvious absurdities, written to appeal to an audience of adolescent boys by writers only slightly older. Although its manufacture would eventually command the most sophisticated resources of the entertainment industry, SF's essential appeal and target audience have not changed. The SF movies that have been most successful—such Top Ten moneymakers as *E.T.*, the *Star Wars* trilogy, *Terminator* 2, and *Independence Day*—have been those that have most scrupulously honored the Boys' Own Adventure formulas of the genre's humble beginnings.

This fact alone does not account for science fiction's position for so long as the leper of literary genres. That can be explained in part by a quick survey of typical SF movie posters of the '50s and '60s, when SF meant low-budget drive-in movies designed to attract male adolescents conflicted about their sexual appetites who thrilled to imagine themselves metamorphosed into teenage werewolves, lustful robots, and other beings with strange kinds of skin, the very sight of whom would cause young starlets to scream and run. For many years, such monster movies were what people had in mind when they dimissed science fiction, with a knowing smile, as the intellectual equivalent of acne—a disfiguring but temporary affliction.

Even today, the schoolmarms among us tend to look down their noses at anything bearing the stigmata of sci fi, though now it would be *Star Trek* they would condescend to. Even then, their contempt would be hedged by a sense that SF is sometimes respectable, for its basic modus operandi has spread through the intellectual environment like a computer virus. Among mainstream writers of both middle- and highbrow who have recently published science-fiction novels, one may cite Doris Lessing, Gore Vidal, Margaret Atwood, Peter Ackroyd, Ira Levin, P. D. James, Paul Theroux, and, preeminently, Michael Crichton.

Since the publication of *The Andromeda Strain* in 1969 until today, as *The Lost World* sets new box office records, Crichton has been, in a commercial sense, the most consistently successful SF writer of the late twentieth century. In large part this is because he has not been labeled *as* an SF writer, and thus unworthy of adult attention. The plots of his best-sellers have featured viral plagues from outer space, lost tribes in darkest Africa, electronic behavioral control, berserk robots, UFOs crashed in the ocean, and dinosaurs terrorizing modern cities—all venerable SF tropes, but somehow not SF when Crichton handles them. Why is this? Because they aren't set on spaceships or other planets but in a plausible present, modified in only the one particular the book focuses on. Even then the novelty he offers is something his audience already half-believes in. When the film of *Jurassic Park* first appeared, all the media hype was designed to make it seem that science was on the brink of resurrecting dinosaurs, for Crichton, and the marketing machinery behind him, realizes that telling a whopper is not enough. People want to *believe* such fictions. Hence, the authenticating "science" in the compound "science fiction," with its implicit guarantee that this dream might come true, as against the surreal or supernatural events of fantasy and fable.

Rocket ships are SF and magic carpets are fantasy, even though those who ride them might be similarly costumed and having almost the same adventures. Stories of time travel are accounted SF, as are tales of telepathy and other psychic powers—this despite the fact that time travel is almost surely an impossibility, and psychic powers belong to the realm of imposture and not science. A few SF writers of a rationalistic bent—so-called hard-core SF writers—have tried to define the genre in such a way as to exclude stories that traffic in scientific impossibilities, but even these writers finally give in to fiction's need for flying carpets in the form of faster-than-light rocket ships, without which SF could not freely venture beyond our own solar system.

Is it then the case that SF is entirely a matter of labeling rather than content? It often would seem so. Some years ago at a PEN Conference in New York City, I was approached by the writer and academic Morris Dickstein, who asked me, in a puzzled way, whether I considered George Orwell's *1984* a work of science fiction—a heresy he'd just encountered. When I assured him that I did, he wrinkled his nose, as though taking a pinch of invisible snuff, and said, "Really!" To Dickstein, an accredited intellectual of Orwell's magnitude could not, by definition, have written science fiction. Never mind that Orwell was writing about a future England, drastically transformed from the England of 1948 but logically extrapolated from existing historical trends, with an appropriate new technology (interactive television). If Orwell wrote it, it was literature, and could not therefore be called science fiction.

It can be galling, for those who have dwelled within the ghetto walls, to be reminded, as Dickstein that day reminded me, that they are not first-class citizens, but there are associated benefits. Just as laboratory rats who are never fed to satiety tend to live longer (albeit hungrier) lives, so science fiction writers, unheeded beyond the ghetto walls, are often uncommonly productive. Survival, for genre writers, depends on productivity—at least a book a year. Isaac Asimov and Robert Heinlein continued at that pace, or better, to the brinks of their graves, and Frederik Pohl and Arthur Clarke, both in their late seventies, are still certifiable workaholics. Those who can't hold to that pace lose their place on the assembly line and are forgotten.

Often, as I came up through the ranks of SF professionals, I had the instructive experience of meeting the SF equivalent of Norma Desmond—once-idolized writers no longer productive but still haunting SF conventions for the sake of the recognition to be wrung from those who could remember reading their books. The late Alfred Bester was the most minatory example. In the '50s he'd written two books, *The Demolished Man* and *The Stars My Destination*, that had secured for him a reputation as the most literary SF writer of his time. Then Bester graduated to a mainstream job as an editor for *Holiday* magazine. His fiction dwindled and lost its edge; at last, after an attempted comeback had come to nothing, he retreated from Manhattan to the boonies. When he died in 1987, he named his bartender the heir to his home and his literary estate.

Bester's mistake was growing up. If the golden age of science fiction is twelve, it follows that SF writers will be successful in proportion as they can maintain the clarity and innocence of wise children. Writers as diverse as Ray Bradbury, Harlan Ellison, Anne McCaffrey, Piers Anthony, and Orson Scott Card all owe a good part of their popularity to their Peter Pannishness. Characteristically, their stories do not pay much heed to those matters of family and career that are the usual concern of mature, responsible adults and the mature, responsible novelists who write for them, like John Updike and Anne Tyler. Many classic novels and stories of the genre are about children of exceptional wisdom and power: A. E. Van Vogt's *Slan*, Theodore Sturgeon's *More Than Human*, Orson Card's *Ender's Game*. In my own golden age, such tales did wonders for my self-esteem, and when I became an SF writer myself, I passed along the torch in more than one book that featured children or youths of preternatural ability.

"Neoteny" is what biologists call it: the retention of larval or immature characteristics in adulthood. It is an essential element of the creative process across the board. Think of the abstract expressionists smearing their giant canvases with oozy oil paints, like divinely inspired kindergartners. Rock stars turning temper tantrums into song. Dancers cavorting in tights and tutus. Art is a kind of play, and those who forget how to be playful are likely to produce art that is ever more mature and responsible and ponderous. Science fiction did not invent the wise-child protagonist. Dickens, Twain, Cather, and Salinger have all produced classics in the same vein. The difference is that for SF that vein has become a main artery. Even when a tale's protagonist is not a legal minor, his or her attitudes, actions, and audience appeal are more likely to be in the spirit of Captain Marvel than of Andrew Marvell.

Science fiction can be neoteric in many ways, from the immortal dopiness of a Flash Gordon movie serial or Ed Wood's *Plan Nine from Outer Space* to the superb dioramas of Arthur Clarke's sagas of the exploration of the solar system, with their detailed depictions of technologies not yet invented and landscapes no man has ever seen. Reading Clarke's SF of the '50s and '60s was like going to a natural history museum that featured spaceships instead of dinosaurs, the future instead of the past. Children and adolescents also have their own distinctive ideas concerning humor, sex, politics, and prose, and their tastes in these matters may strike older readers as sophomoric, gauche, ill informed, or just dead wrong. Conversely, the young have a way of noticing that good manners can be oppressive, that the past is often irrelevant, and that emperors are sometimes naked. In short, the young are not lesser beings; they're just different.

One of the most salient differences is their relationship to the past and the future. Grown-ups have experienced at least a bit of the past; children must imagine it just as they (and grown-ups, too) must imagine the future. A few years can make a profound difference in one's zeitgeist, our sense of how the present meshes with history. I was born in 1940 and remained unconscious of the greatest event of the century as it was happening. Anyone only a few years older would have incorporated the terrors and triumphs of World War II into their soul's fiber. By the same token, for those born after 1970, Hiroshima and the *Apollo* moon landing are not epochal events but ancient history.

For the young, all history is ancient, something that must be learned in school, like the multiplication tables. It is the warehouse in which the culture stores its myths, and while some of the figures in these myths—cowboys, knights in armor, pirates, and other violent offenders with distinctive wardrobes—are exciting enough not to require enforced attendance, as at Sunday school, most historical personages have been so thoroughly denatured and sanitized that the past presented to us in the classroom is justly regarded with indifference or suspicion. The story about George Washington and the cherry tree was not only an invention, it was a plagiarism, borrowed by Washington's biographer, Parson Weems, from an earlier work of cheap fiction. History, as Henry Ford observed, is bunk.

The future is another matter entirely. The past by definition is over and done with, a photo album filled with pictures of dead people and buildings that have been torn down. The future, however, like Christmas, is waiting for us to arrive. The young know they're going to go there, and so they furnish it with their wishes. In the 1920s and 1930s, when American SF was aborning, its menu of future wonders was a national letter to Santa Claus listing the toys that boys like best—invincible weapons and impressive means of transportation. When the future began to arrive, in the '50s and '60s—that is, when the dreams of the SF magazines began to be translated into the physical realities of the mature consumer culture by a generation of designers and engineers who'd come of age in the pulp SF era—cars were streamlined to resemble rocket ships. In fact, the car was revealed as the secret meaning of the rocket ship—a symbol, at gut level, of absolute physical autonomy.

The complex equation of car and rocket ship epitomizes the relationship between SF and the surrounding culture. There is no more persuasive example of the power of "creative visualization" than the way the rocket-ship daydreams of the early twentieth century evolved into NASA's hardware. Between them, the SF pulps and such kindred publications as *Popular Science* and *Mechanix Illustrated*, in which the new technologies of the burgeoning industrial state were set forth in terms any bright twelve-year-old could understand, produced the blueprints for the building of the land of Oz and the home of Ozzie and Harriet.

Inevitably there will be discrepancies between any set of blueprints and what finally gets built. Some preliminary sketches are transparent fictions, which the architect never intended to be taken seriously. There *won't* be a heliport on every skyscraper, despite all the great illustrations in the old pulps and the set designs in Luc Besson's *The Fifth Element*. The traffic control problems would be too great. On the other hand, within a single generation, between 1950 and 1970, jumbo jets did supplant trains and buses in providing long-distance mass transportation, so that at least in an allegorical sense, anyone who keeps track of Frequent Flyer bonuses does have a home heliport.

This "allegorical" sense is often SF's acutest receptor. SF writers often score prophetic bull's-eyes while getting all the details wrong, as Orwell did in *1984*. They can also be uncannily accurate at microprediction and still miss the big picture. Aldous Huxley's *Brave New World* seems more prophetic every decade. Technology keeps getting closer to creating true test tube babies, and human cloning looms ahead. Today's blockbuster movies are as mindless, thrilling, and overtly pornographic as his feelies. The Deltas and Gammas of the lower classes are sustained by sex, sports, and drugs, just as in his novel. But in our reality drugs are illegal, the streets are dangerous, and human nature has managed to defeat most efforts at social programming. The hedonic utopia that Huxley saw latent in the America of the 1930s remains a latency.

This book provides a key to the allegories of science fiction and chronicles the genre's impact on American, and, eventually, global, culture. That impact is not always so straightforwardly causal, as in the case of the rocket ship. Some of SF's most enduring and recurrent images, such as its obsession with robots that run amok, have been very wide of the prophetic mark. We live in a world swarming with robots, which we generally take for granted or are blind to. They pilot planes, operate elevators, cook dinner, build cars, record TV shows when we can't do it ourselves, and rarely run amok, though they may misfunction. These "robotic" byproducts of the computer age, although they have transformed our lives in myriad ways, lack the anthropomorphic robot glamour of "iron men," the pathos of being intelligent but soulless, and the high drama of rebellion against one's creator. The natural penchant of any storyteller for high drama over mere logic led even such capable extrapolators as Isaac Asimov to write about robots that were "almost human" and thereby to fail to foresee the cybernetic future until it was already upon them.

This does not mean, however, that SF's long-time preoccupation with robots amounts to no more than a failure of foresight. The robots of the sci-fi imagination have a different significance, which can be seen in their first appearance on the literary stage, in the 1920 play *R.U.R.* by Czech author Karel Capek. Capek coined the word *robot*, from a Czech root meaning "serf labor." His robots are a nightmare vision of the proletariat seen through middle-class eyes at the historical moment of the first Bolshevik success in Russia. They are manufactured, and therefore property (as Russian's serfs had been). Being the cheapest possible source of labor, they allow the privileged humans of the play to enjoy lives of sybaritic splendor until the moment the robots, realizing their own strength, rebel and wipe out their manufacturers. Capek's sympathies waver between indignation on behalf of the exploited robots (which sometimes seem to have souls) and fear of the impending day of judgment that will bring middle-class privilege to an end. The SF device of substituting robots for human workers allowed Capek to express, in the telegraphy of allegory, the moral truth that the industrial system treated human laborers as though they were machines, sowing thereby the seeds of an inevitable and just rebellion.

Many plays and novels and tracts have been written to express the truth that the working class are people just like "us": Emile Zola's *Germinal,* George Bernard Shaw's *Pygmalion,* John Steinbeck's *The Grapes of Wrath,* and others. What *R.U.R.* is able to communicate that none of those other works can permit themselves to suggest is the ethically incorrect horror of a proletarian revolution that would put the rabble in charge. Capek is telling us something about ourselves we would rather not know: that deep down we don't believe in the humanity of those whose labor we exploit. And not just the proles in sweatshops and factories, for in Capek's time virtually every middle-class household had its own staff of "robots" in the form of cooks, maids, and scullions.

This vision of the essential inequity of all servant-master relationships eventually supplanted a more familiar and benign view of the nineteenth century: that servants, because they lived in the same house and shared some of its comforts, felt a family-like loyalty to those they served. This is the myth embodied in *Gone with the Wind,* in the TV series *Upstairs, Downstairs,* and in P. G. Wodehouse's fables of Jeeves and Bertie Wooster. The myth can be controverted without resorting to the distancing devices of SF, but usually in such countermyths the servant enjoys an underdog victory, as in Beaumarchais' Figaro plays or and J. M. Barrie's *The Admirable Crichton,* in which a butler, shipwrecked on a desert island with his employers, becomes their overlord by virtue of his greater native endowments. Only if workers and servants can be shown to be something other than human is it possible to express, guiltlessly, a disdain for, as milord would have it, the canaille.

This uneasiness with regard to keeping servants, and keeping them in their place, became especially acute in the United States, both because it was, from its inception, a democracy in which all men were supposed to be equal, and because at the very dawn of the modern industrial era, it went through the traumatizing experience of the Civil War, which was fought over the issue of slavery. *R.U.R.* appeared on Broadway in 1922, only two years after its Czech premiere, and America at once assimilated the idea of the robot. Only a few years later, the poet Kenneth Fearing was filled with identical forebodings:

> Only Steve the side-show robot, knows content; only Steve, the mechanical man in love with a photo-electric beam remains aloof; only Steve, who sits and smokes or stands in salute, is secure;
>
> Steve, whose shoebutton eyes are blind to terror, whose painted ears are deaf to appeal, whose welded breast will never be slashed by bullets, whose armature soul can hold no fear.[1]

The most terrible fears are often those we are not allowed to express and which must therefore be displaced to a permitted bogey. The witch who enslaves and imprisons Hansel and Gretel is not, heaven forfend, their mother, or even their wicked stepmother, but an Other. And the Steve whom Fearing fears is not some stevedore working the New York docks (for Fearing accounted himself a leftist and first won recognition in the pages of magazines like *New Masses*); rather he is an Other, not human, a robot whose blind eyes and deaf ears even a leftist might regard with consternation.

The robot has been employed for as wide a variety of dramatic purposes as there are Others to be worried about. We will encounter many other Others in these pages: aliens with green skin instead of green cards; body-snatching aliens inhabiting our neighbors' borrowed flesh; androids and cyborgs (robots in disguise); Artificial Intelligences, or AIs (robots who have shuffled off all mortal coils, metal *and* flesh); assorted gods and demigods (most notably Valentine Michael Smith, the hero of Robert Heinlein's *Stranger in a Strange Land* and Charles Manson's fatal role model); and, last but not least, those most Significant Others, women.

Women were usually neglected by SF writers of the earliest era, except when there was a need for their Lois Lane capacity as damsels in distress. However, ever since *R.U.R.*, there have been some notable female robots. Lester del Rey, a minor SF writer who would later give his name to a major SF imprint, wrote a story in 1938, "Helen O'Loy," that is a classic example of "Golden Age" (i.e., prepubescent) sexual psychology, in which two chums share the love of a mail-order, ready-to-assemble mechanical bride. "She was beautiful, a dream in spun plastics and metals, something Keats might have seen dimly when he wrote his sonnet." Further, "Helen was a good cook; in fact she was a genius, with all the good points of a woman and a mech combined."[2] That's about it, in terms of characterization, but the seed had been planted that would become, in 1972, Ira Levin's *The Stepford Wives*, and, in 1974, a classic movie that put a feminist spin on del Rey's basic equation, Housewife = Robot.

Housewives are, after all, the last domestic servants—or so it seemed in the early '70s, at the zenith of the American consumer culture, when every kitchen had its little battalion of labor-saving appliances. The Stepford wives were not scullions but rather, like del Rey's Helen O'Loy, succubae catering to the hedonist requirements of their lords and masters. Like Helen, they could pass as human. Their robot nature was a secret they shared with their husbands—at least, until the movie made their name a byword for marriage as the most intimate form of alienation.

From the start, science fiction has had a double nature. At its crudest it is the ringmaster for monsters from the Id, bubbling with crude wish-fulfilling fantasies, as in "Helen O'Loy." But such fantasies can be very potent. They will capture the attention not only of a naive audience but of all those alert to such fiction's primal meaning: to grown-up writers like Ira Levin or Margaret Atwood who can recognize their own features in the comic book grotesqueries of naive sci fi and who then do their own sophisticated recensions of the crude originals.

This dialogic process has been going on so long, among so many different writers, that the confusion of realms between highbrow and low, between naive and knowing, has become a cultural fait accompli. The machineries of Hollywood pour millions of dollars into creating ever more artful re-creations of comic book heroes, from Superman to Caspar the Friendly Ghost, while the hacks who write *Star Trek* tie-ins and Marvel Comics take their cues, as often as not, from the writings of Michel Foucault and Camille Paglia. Newt Gingrich has a stable of collaborators for both his fiction and nonfiction who are seasoned sci-fi professionals.

In short, science fiction has come to permeate our culture in ways both trivial and/or profound, obvious and/or insidious. And its effects have not been limited to the sphere of "culture," in the narrow sense of one art form's influencing others. The influence of science fiction, as we shall witness abundantly in the pages ahead, can be felt in such diverse realms as industrial design and marketing, military strategy, sexual mores, foreign policy, and practical epistemology—in other words, our basic sense of what is real and what isn't.

It is my contention that some of the most remarkable features of the present historical moment have their roots in a way of thinking that we have learned from science fiction—to wit: the razing of the Berlin wall; the rise of millennial cults with homicidal agendas; Oliver North's testimony before Congress and his campaign for a Senate seat; Madonna's wardrobe and Sinead O'Connor's hair style; celebrity murder trials; compassion "burnout" for refugees in Rwanda; the deaths of the *Challenger* astronauts; toxic waste cover-ups; and much too much more.

In 1938, the year that "Helen O'Loy" appeared, the poet Delmore Schwartz, age twenty-four (two years younger than Lester del Rey), published his first collection, *In Dreams Begin Responsibilities*. That title alone secured his immortality. This book is an amplification of that pregnant truth.

A few words about my own connections with SF.

If American science fiction begins with the first issue of Hugo Gernsbach's *Amazing Stories* in 1926, then I have been a professional science-fiction writer for just about half the time the genre has been in existence, and I've been reading the stuff for two-thirds of that time.

Of the one hundred to two hundred writers of the genre whose bylines are likely to register among avid readers of SF, I have met a majority. In 1980 and 1983 the British SF writer Charles Platt brought out two anthologies of interviews with well-known SF writers, *Dream Makers* and *Dream Makers*, Vol. II. Of the thirty writers in the first book, I'd met all but two; of the twenty-eight in Volume II, I'd met nineteen. I'd roomed with some, dined with most, had business dealings with many, reviewed their work and been reviewed by them, debated with some in public, and gossiped with all of them about the others.

In my experience moving from one literary frogpond to another, I have never encountered a group of writers so intensely and intricately interconnected as the SF community. Poetry comes closest, but poetry is balkanized into dozens of hostile or indifferent clans. The various bands of the multicultural rainbow tend to be separatist both socially and aesthetically. When old and young do intermingle, it is in the institutional setting of a classroom or a summer workshop. But the chief difference is this: poets have a few centuries of other poets' work to catch up on. A poet can avoid reading contemporaries altogether and still read widely, deeply, and relevantly.

By contrast, most of the science fiction that is worth reading has been written by the writers I've met—some of whom, like Theodore Sturgeon and Robert Heinlein, began publishing in the late '30s. As recently as 1981, when I wrote a foreword for the SF volume of *Gale's Dictionary of Literary Biography*, I could declare that all the great SF writers were essentially contemporaneous, alive and well, and merrily cross-pollinating across the usual gaps of age, gender, and ideology. Since then, Robert Heinlein, Isaac Asimov, and nine more of the sixty SF writers Platt interviewed have died. Even so, there are few other fields of endeavor—quantum mechanics, computer design, genetic engineering—in which so large a proportion of its most illustrious figures still figure in reference books with only a single date in the parentheses after their name. And there are even fewer fields that have had so brief a history in proportion to the extent of their cultural impact.

This book is about that impact. It is not a literary history. Some of the science fiction I value most highly as literature—books by John Crowley, Gene Wolfe, and Paul Park—is dealt with only in passing, because the impact has been slight. These works are admired by discerning readers within the field, but being inimitable, they have not been imitated. By contrast, some of the most influential and widely imitated writers in the field—Asimov, Heinlein, Herbert, Pournelle, Card—vaunt themselves on their artlessness and lack of literary polish, or at least that is the John-Wayne persona they affect. They are simple "tellers of tales."

I also have no intention of "debunking" science fiction, though I'm sure I will give offense to many of the writers discussed (and their admirers)—and even graver offense to those not discussed at all. That much SF is written for the very young and/or the uninstructed is a fact of publishing demographics. Indeed, the SF that reaches the largest audiences—earns the biggest grosses, and establishes its archetypes most firmly in the collective mind not just of the nation but of the globe—is not published at all; it is broadcast over TV and screened in movie theaters. One could dismiss such work as being aimed at the "lowest common denominator," but one could dismiss the Gospels on the same grounds. Blessed are the poor in spirit? Well, then, blessed are the Trekkies, too. Theirs is the Kingdom of Heaven.

Having lived in the world of SF so long and having known so many of the players, it would be a false modesty to exclude my own personal witnessings from this account. That said, I must add that this book is not a memoir, nor an apologia for some one set of aesthetic principles, my own. Indeed, there is no such set, for my taste in SF has been indiscriminate. At one time or another I loved it all. I have doted on E.C. comic books; on Asimov serials in *Astounding* (which inspired my own first space opera plots, scribbled on nickel tablets at age twelve); on the satirical novels of Kornbluth and Pohl (role models for my first, unfinished SF novel, begun and aborted when I was twenty-three); by hundreds of other writers, arrant hacks and unsung geniuses. As time went on, some of those enthusiasms diminished, some held steady, and others formed. In my years as a Young Turk in the late '60s, I burned with the intolerance of a true faith, the New Wave, which was to elevate SF to its true potential as the heir of Joyce and Kafka, Beckett and Genet. Now, with the benefit of distance, I can afford to be tolerant—and hope to be objective.

Chapter 1

THE RIGHT TO LIE

America is a nation of liars, and for that reason science fiction has a special claim to be our national literature, as the art form best adapted to telling the lies we like to hear and to pretend we believe.

It has been said of Cretans that they were all liars, and we can assume, from its proscription in the Decalogue, that lying was not unknown in Mosaic times. What distinguishes American liars from those of earlier times and other nations is that the perfected American liar does not feel himself to be disgraced by his lies, even when he is caught in them. Indeed, the bolder the lie and the more brazenly imposed on the public, the more admiration the liar is accorded.

The first American hero to be celebrated for his wily ways is a folk spirit native to the continent, Coyote. Among his lineal descendants in the realm of fiction one may number Joel Chandler Harris's Br'er Rabbit, Herman Melville's Confidence Man, Mark Twain's Tom Sawyer, and Abigail Williams in Arthur Miller's *The Crucible,* the Puritain maiden whose lies give rise to the Salem witch trials.[1] What sets such American tricksters apart from those of other cultures is the degree to which they solicit our admiration. I can remember my father's reading aloud the opening chapters of *The Adventures of Tom Sawyer* and the delight we shared at the way Tom hustles his friends into whitewashing a thirty-yard-long, nine-foot-high board fence.

Before addressing the SF component of this issue, let me offer a short anthology of righteous lies from the past forty years of American history by way of suggesting the dimensions of the nation's Great White Fence. The first great lie of the post–World War II era, and the foundational whopper of the Cold War, was President Eisenhower's scout's-honor insistence, in 1960, that the U-2 shot down over the Soviet Union was not on a spying mission. That pro forma diplomatic fiction became a booby trap when the pilot, Francis Gary Powers, whom Eisenhower had presumed dead, was produced alive and stood trial for spying. Sisela Bok declares that "this lie was one of the crucial turning points in the spiraling loss of confidence by U.S. citizens in the word of their leaders."[2]

The Vietnam War offered Americans a more extensive lesson in their government's complacent disregard for inconvenient truths. In *The First Casualty*, his history of war reporting, Phillip Knightley writes, concerning Vietnam:

> In the early years of the American involvement, the administration misled Washington correspondents to such an extent that many an editor, unable to reconcile what his man in Saigon was reporting with what his man in Washington told him, preferred to use the official version. John Shaw, a *Time* correspondent in Vietnam . . . says, "for years the press corps in Vietnam was undermined by the White House and the Pentagon. . . . Yet the Pentagon Papers proved to the hilt that what the correspondents in Saigon had been sending was true.[3]

Knightley contends that compared to earlier wars of the modern era, the press (though not the government) had a good track record for honesty. "But," he admits, "this is not saying a lot. . . . With a million-dollar corps of correspondents in Vietnam the war in Cambodia was kept hidden for a year."[4]

But it was Watergate that made clear even to the most trusting and credulous of citizens that Presidents, their advisers, and anyone within distance of a bribe have as little regard for the truth as Richard III. Nixon had lied so successfully for so long about matters of such consequence that for the entire first year of the scandal, he refused to believe his robes of office would not protect him. In his steadfast denials, which he persisted in even after resigning in disgrace, he set an example of the Liarly Sublime that has never since been bettered for sheer brass.

The final establishment of lying *as a right*—a right that is specifically God given—was the work of Marine Corps paragon, presidential adviser, and 1994 Republican candidate for the U.S. Senate, Oliver North. In July 1987, North had been called to testify before the Senate concerning the White House's involvement in trading arms to Iran in exchange for the release of American hostages, the diversion of those illegal funds to assist (in defiance of Congress) the contras in Nicaragua, and his own perjuries with respect to these operations, which he had superintended. North's biographer, Ben Bradlee, Jr., begins his summing up of all the lies that were Oliver North's life:

> Aside from North's admitted lying to Congress about the Contras, his admitted lying to the Iranians, his admitted falsifying of the Iran initiative chronology, his admitted shredding of documents and his admitted lying to various Administration officials as the Iran-Contra affair unraveled in November of 1986, there are stories, statements or claims that he has made to various people while at the NSC [National Security Council] that are either untrue strongly denied, or unconfirmable and thought to be untrue.[5]

Bradlee recounts a round dozen of North's wilder whoppers, which include: a variety of self-promoting tales about confidences and private prayer meetings shared with the President; a tale of derring-do concerning his rescue of wounded contra soldiers (who later died, alas) as he piloted a plane through enemy machine-gun fire; his service in Angola and in Argentina during the Falklands War, and strategic tête-à-têtes with Israel Defense Minister Ariel Sharon just before the Israeli invasion of Lebanon (all three stories pure fabulation); and his dog's death by poisoning—"presumably by those whom he said had been threatening his life." (A neighbor insisted that the dog died of cancer and old age.) The list is extensive enough to suggest that North's penchant for lying exceeded the merely strategic and expedient and amounted to pathology, and this is confirmed by the testimony of even reputed friends.

So artful was North's performance before the Senate that soon a good deal of the country had adopted the same attitude. "No one has captured the American public like Ollie North," opined a Chicago restaurateur. "Even when he's not telling the truth he's beautiful. The guy is so charming." Then columnist and future presidential candidate Patrick Buchanan lauded North as "a patriotic son of the republic who, confronted with a grave moral dilemma—whether to betray his comrades and cause, or to deceive members of Congress—chose the lesser of two evils, the path of honor. It was magnificent."

There must be two parties to a successful lie: the one who tells it and the one taken in. The motive of the teller is seldom difficult to discern, though it may be complex. In North's case one can scent self-advantage, a desire for applause, a certain amount of rational fear, and, not least, an inveterate delight in his own con-artistry. Surely a good deal of North's success was due to the TV audience's collusive admiration for the man's brass. Like Buchanan, they knew he was lying, but he lied so well; it was magnificent. This was the era, after all, of a President who had been caught again and again in evasions and fabulations. But people didn't care. Indeed, they applauded both men's acting skills—the catch in the throat, the twinkle in the eye, the scout's-honor sincerity. TV critic Tom Shales favorably compared North's debut at the Senate hearings to Burt Lancaster's performance in *Seven Days in May,* in which Lancaster plays a general planning a military *coup d'état.*

I've said America is a nation of liars. A politer way of putting it is that we are a nation of would-be actors. No other culture has ever been so drenched in make-believe. Children spend more time watching television than going to school, and most of what they watch is fiction. In school they are taught to read novels. Actors are national celebrities, and show business is widely recognized as a metaphor for the conduct of life. A smile on one's face and a shine on one's shoes are the simple prerequisites for success in a world of self-made men. We could make believe; it's only a paper moon; let's go on with the show.

So where does that leave the vast majority of us whose role is only to watch the stars, to applaud, to believe, to vote? There are, admittedly, a great many who do not succumb to the blandishments of the entertainment industry and who can distinguish between performances and principles. But by and large our media stars are admired. They are spoken of often as "role models," that is to say, templates on which we can form our own social personae. Even their sins are trend setting.

And so, if Reagan and North may lie with impunity, why shouldn't we all be allowed the same latitude? Few private citizens can take refuge behind "national security" as an all-purpose excuse for self-aggrandizement, but there are any number of worthy causes that an expedient lie can take shelter in. Take the case of Tawana Brawley, a black teenager who claimed, late in 1987, to have been attacked and sexually abused for four days by a gang of white cops, including (according to statements later made by her attorneys, Alton Maddox and C. Vernon Mason), an assistant district attorney and another local law officer, Harry Crist, Jr., who had committed suicide shortly after Brawley's story hit the news and so could not deny the lawyers' allegations. Brawley was taken up as a martyr by black activists. She and her mother, who had colluded in her fabrications, became protégés of the black demagogue Rev. Al Sharpton, who took them out of state and beyond the summons of the grand jury investigating her alleged rape—and thereby spared the hard choice between perjury and admitting to the shameful truth; to wit, that she had fabricated the whole story, smearing herself with dog feces, scrawling "Nigger" and "KKK" on her body with charcoal, and scorching the crotch of the jeans she'd been wearing. Yet to this day, nine years later, despite a grand jury report that presents the great mass of evidence that shows Tawana was lying, the Rev. Al Sharpton equivocates about her probity, on the grounds that even if this particular crime did not take place, others like it have.

Since Tawana's time, allegations of sexual abuse have become epidemic, but later liars have learned from Tawana's example not to tell lies that can so easily be disproved. Of great usefulness in this regard has been the Recovered Memory Syndrome, in either its simple form or in combination with fantasies of ritual satanic abuse. A catalogue of only the most celebrated cases of recent years would take pages and would be a work of supererogation, since most large bookstores now have entire sections devoted to the phenomenon.

That child sexual abuse occurs cannot be denied, but even when it is reported shortly after it is alleged to have happened, a certain skepticism is called for. The supposition that children are more likely to be truthful than their elders is unwarranted, especially in a culture of liars. One cautionary tale was a recent case in Chicago, in which three sisters, ages ten, eleven, and twelve, alleged that they had been the victims, at their father's hands, of four years of sexual assault, beatings, drug injections, and meals of fried rats and boiled roaches. The last detail secured the case national attention, but it probably was the fatal, over-the-top flaw that saved the girls' parents from prosecution.

Generally children are held as little accountable for their lies as for their more overt misdemeanors and felonies. In theory they are innocents a priori. Many grown-ups feel that they are entitled to similar license, and for them the Recovered Memory movement has been a godsend. Are you obese? Underweight? Anxious? Frigid? Sexually hyperactive? Then, according to countless self-help books, you were probably a victim of childhood sexual abuse but have repressed the memory of it.[6] These memories are to be recovered by means of group therapy, hypnosis, and massage.

How this process enables those making allegations and even those falsely accused to fabricate scenarios and then *believe* in their own inventions has been described by Lawrence Wright in *Remembering Satan* (1994), the stranger-than-fiction account of a man railroaded into prison after his daughters had charged him with ritual satanic abuse. Not only were the two daughters and his wife able to "recover" memories of sins that had never been committed, even their victim cooperated in producing more specious "memories." As to the psychological theories that are the foundation of the movement, Frederick Crews has done a most thorough demolition job in two essays originally published in the *New York Review of Books* and reprinted, along with the outraged correspondence they provoked, in *The Memory Wars: Freud's Legacy in Dispute* (1996). Crews contends that there is as little intellectual and evidentiary substance in the theory and practice of psychoanalysis as in the most blatantly fantastical claims of believers in ritual satanic abuse. Indeed, the latter, Crews urges, is the devolved and déclassé descendant of the former.

I would agree with Crews, with this further suggestion: that science fiction has been an essential element in the transmission of Freud's original theories and their adaptation to the needs of today's talk show audiences. There were actually two separate SF conduits. The first was the debased Freudianism of SF writer L. Ron Hubbard, who introduced the pseudoscience of Dianetics (aka the "religion" of Scientology) in the May 1950 issue of *Astounding Science Fiction.* The second and more direct route is that typified by Whitley Strieber, a writer of horror novels who claims to have been abducted and sexually abused by aliens at periodic intervals throughout his life, a fate subsequently shared by his son, then age seven. Strieber's books on this subject, *Communion: A True Story* and *Transformation: The Breakthrough,* remain notable for being the only such books by an already established professional author, for which distinction Strieber received a whopping million-dollar advance for *Communion.*[7]

The symptomatic relationship between L. Ron Hubbard and science fiction will be examined at greater length in Chapter 7, on SF and religion. Strieber's and other "abductees'" memoirs of their UFO experiences might also be considered from that higher vantage—had they been received by the media and the general public with the solemnity and immunity from skeptical examination that is tacitly accorded to officially recognized religions. Happily, though Strieber had a commercial success with *Communion,* his effort to form a quasi-religious cult of alien abductees did not attain orbital velocity, and so more than a decade after his alleged abduction on the night after Christmas 1985, *Communion* has become a part of the history of pop culture, not of religion.

Whitley was not the first UFO hoaxer, though he has been, to date, the most audacious and has turned the best profit from his fancies. The first rash of purported sightings was the Great Airship Mystery of 1896, when an armada of cigar-shaped airships with winglike sails crossed the skies of America. These protodirigibles clearly were an expression of imaginative enthusiasm for the dawning era of heavier-than-air flight, and once that era had commenced in fact, such fictions disappeared.

Then, at the dawn of the atomic era, came the flying saucers. In 1947 Kenneth Arnold reported nine disk-shaped objects sporting about Mount Rainier, and before the year was out, 850 other "flying saucer" sightings had been reported in the press. From the first, those who credited flying saucer sightings assumed them to be of extraterrestrial origin—which is another way of saying that they were the progeny of science fiction. Orson Welles's radio dramatization of H. G. Wells's *The War of the Worlds* had provoked a panic among credulous listeners in 1938—proof, if any were needed, of a large audience of potential believers.

But strange lights in the night sky are not enough. Contact was inevitable, and on November 20, 1952, George Adamski, a penny-ante guru already in the flying saucer business, lecturing on the subject and selling his own UFO photos, had his first tête-à-tête with a Venusian named Orthon, who explained by dumb show and telepathy that his saucer was powered by Earth's magnetism. After some brief instruction in English, Orthon was able to express ("Boom, boom!") the alarm of peace-loving extraterrestrials at atomic testing and the prospect of nuclear war. (Only a year earlier in the 1951 sci-fi movie, *The Day the Earth Stood Still*, a flying saucer lands in Washington, D.C., to deliver the same message. A coincidence?)

Adamski's ghost-written account of his contact with Orthon, *Flying Saucers Have Landed* (1953) and its sequels, *Inside the Spaceships* (1955) and *Flying Saucers Farewell* (1961), had already fallen into disfavor among saucer buffs by the time of his death in 1965. Adamski's style of fakery and his narrative gifts were too primitive for the new breed of UFOlogists, and he had alienated even the faithful readers of his *Cosmic Bulletin*, as his emphasis shifted to mysticism and psychic phenomena. Originally Adamski had declared psychics to be in cahoots with the world banking interests, the "Silence Group," which acted to suppress information about UFOs. Now he seemed to be saying that his contacts with the space people and his trips to the dark side of the moon might be nothing but visionary experiences.

Adamski's equivocations in this regard were to become a regular feature of the more high-toned UFO "abductees" and their chroniclers of a later generation. Well before Strieber hit the scene, Dr. J. Allen Hynek, dubbed by *Newsweek* as "the Galileo of UFOlogy," gave grudging credence to the first abductee narrative, that of Betty and Barney Hill.[8] But by 1982 Hynek was backing away from the received wisdom that UFOs were of extraterrestrial origin: "The enigma to which [Hynek] had dedicated his career remained inscrutable and unacceptable to the scientific community. Hynek submitted that perhaps UFOs were part of a parallel reality, slipping in and out of sequence with our own. This was a hypothesis that obviously pained him as an empirical scientist. Yet, after thirty years of interviewing witnesses and investigating sighting reports, radar contacts, and physical traces of saucer landings no other hypothesis seemed to make sense to him."[9]

Parallel universes are another trope borrowed from the repertory of science fiction. They are a marvelous convenience for authors who want to fantasticate at a high rpm without having to offer a rational explanation for the wonders they evoke. In a parallel universe, magic is usually the operative technology, as per SF fandoms' motto, "Reality is a crutch." Well before Hynek had adopted this all-purpose escape clause, dozens of SF writers had proposed the same explanation for the persistent unverifiability of all UFO phenomena. The cleverest such confection has been *Miracle Visitors* (1978) by British SF writer Ian Watson.

Unverifiability is for UFOlogists what deniability was for Nixon and the Watergate conspirators. Without it, they would be attested perjurers. Accordingly, Strieber and Dr. John E. Mack (the Harvard Medical School Professor who drew flack from his colleagues for his UFOlogical "research") carefully lard their abduction narratives with canny disclaimers. Not for them Adamski's crudely faked photos of UFOs and his footprint castings of aliens who always manage to elude the camera's lens, despite decades of "close encounters." No, the aliens are ineffable and unknowable, and their caprices are no more to be questioned than the koans of a Zen master. In one of his many declarations of independence from rational scrutiny, Strieber declares:

> Whomever or whatever the visitors are, their activities go far beyond a mere study of mankind. They are involved with us on very deep levels, playing in the band of dream, weaving imagination and reality together until they begin to seem what they probably are—different aspects of a single continuum. To really begin to perceive the visitors adequately it is going to be necessary to invent a new discipline of vision, one that combines the mystic's freedom of imagination with the substantial intellectual rigor of the scientist.[10]

In a culture of liars, it is considered bad form to call to account lies reckoned to be harmless. In a telephone survey conducted by the Leo Burnett ad agency, 91 percent of 505 people surveyed confessed that they regularly don't tell the truth.[11] "People," the *New York Times* writer explained, "are more accepting than ever of exaggerations, falsifications, fabrications, misstatements, misrepresentations, gloss-overs, quibbles, concoctions, equivocations, shuffles, prevarications, trims and truth colored and varnished. They even encourage their children to do it by praising them for using their imaginations."

The *Times* itself evinced a similar delicacy when it had to deal with *Communion* in its Sunday book review section, urging its reviewer, Gregory Benford, to adopt a more respectful and accommodating tone. Benford, a noted physicist and an accomplished SF writer, bit his tongue and trimmed his first draft.[12] And so, despite protests over its appearance on the "Non-Fiction" side of the *Times'* best-seller list, *Communion* was accorded respectful attention in the nation's main journal of record, where Strieber is quoted to this effect: "I cannot say, in all truth, that I am certain the visitors are present as entities entirely independent of their observers. Nor can I say that I do not think they are here at all." He cannot say that for a very good reason: if he did, he wouldn't have a book contract.

I reviewed *Communion* for the *Nation*, where I was not under the dueling-code restraints imposed by the *Times* and could freely express my opinion of Strieber's enterprise.[13] Better than that, I was in possession of a smoking gun, for Strieber makes much of his own naiveté regarding earlier UFO testimonies: "I did not believe in UFOs before this happened. And I would have laughed in the face of anybody who claimed contact." He maintained that until he'd been impelled by his own experience to examine other UFO literature, he had taken no interest in such matters. If he *had* read widely in the literature, the striking resemblance between his own UFO experiences and that recorded by others could be ascribed to imitation. But if, as he claimed, he was innocent of such knowledge, then such a correspondence must be seen as a confirmation that Something Is Happening.

Whitley, as it turned out, had left a significant paper trail in this regard: a story, "Pain," that appeared in a 1986 hardcover anthology of horror stories, *Cutting Edge*. "Pain" is a remarkable prefiguration of *Communion*'s distinctive addition of S&M themes to the traditional UFO mixture-as-before. Here is the moment in *Communion* when Strieber reveals how he was raped by aliens:

> [Aboard the saucer] the next thing I knew I was being shown an enormous and extremely ugly object, gray and scaly, with a sort of network of wires on the end. It was a least a foot long, narrow and triangular

> in structure. They inserted this thing into my rectum. It seemed to swarm into me as if it had a life of its own. Apparently its purpose was to take samples, possibly of fecal matter, but at the time I had the impression that I was being raped, and for the first time I felt anger.

In "Pain" the narrator, a novelist like Strieber, tells of his besotted passion for a professional dominatrix, who belongs to an ancient, alien race that had fed on human pain throughout history. They were in charge of the Roman Empire, arranged the Holocaust, assassinated Kennedy, and now their agent, cruel Janet O'Reilly, puts Strieber's hero through a standard bondage-and-domination scenario.

The textual parallels between "Pain" and *Communion* are extensive. Could it be that Strieber, having made the imaginative equation between the "archetypal abduction experience" and the ritual protocols of bondage and domination, realized he'd hit a vein of ore untapped by previous UFOlogists? Strieber's alternative explanation is that the story represents the first surfacing of memories repressed by the aliens, who had given Strieber a similar hazing only days before "Pain" was written.

Strieber's reponses to those who dare to regard his UFO testimonies as other than hard fact or celestial vision have been so passionate that many skeptics are inclined to suppose him sincerely self-deluded. Philip J. Klass, our leading UFO debunker, maintains that Strieber probably suffers from TLE, temporal lobe epilepsy, "a transient phenomenon of the temporal lobe of the brain that causes "vivid hallucinations that are often associated with powerful odors [which Strieber reports during some of his UFOnaut encounters]. . . . People with [TLE] tend to be verbal and philosophical and to lack a sense of humor."[14] If Klass is correct, Strieber believes his own lies. But while Betty and Barney Hill (who let five years lapse between their "abduction" and as-told-to publication) may have been sincerely deluded, Strieber has too visibly and systematically worked to cover his own tracks for such a charitable interpretation to be accepted.

Ever resourceful, Strieber has resorted to another SF trope to explain his penchant for telling lies. His aliens, when they are not probing his nether orifices, have been implanting *false memories*. In an interview with Paul Gagne, Strieber described how he'd almost been a victim of the multiple murderer Charles Whitman, when he opened fire from a tower on the campus of the University of Texas. In Whitley's anguished words,

> I was right in the middle of it. I ended up hiding under a little retaining wall about 2 1/2' high with another person. Everyone near us was shot—not all killed, but shot. As I lay under that wall with this other man right there, a woman suddenly began to scream about ten feet away from us. She was terribly injured. She had been shot in the stomach and she was wailing and bellowing, scrabbling along the ground with blood coming out of her. I was going to run to her when this other man jumped up, and the moment he did, Whitman shot the top of his head off. He had, of course, shot the woman in the stomach for the purpose of getting us to come out from where we were hiding. He was just waiting there with this gun. I didn't move. That has haunted me all my life.

Yet in *Communion* Whitley declares that "for years I have told of being present when Charles Whitman went on his shooting spree from the tower in 1966. But I wasn't there."

Those who've seen Arnold Schwarzenegger in *Total Recall* (1990) will be familiar with the notion of implanted memories. Although Whitley's book antedates that movie, the Philip Dick story on which it was based, "We Can Remember It for You Wholesale," appeared in 1966, and his classic novel *Time Out of Joint* (1959) deploys essentially the same idea. And what a convenience that idea is for any UFO abductee who might find himself mired in provable untruths. This, even more than his equation of abduction with the frissons of sadomasochism, has been Strieber's greatest legacy to the traditions of UFOlogy: nothing one has said or written can be used in evidence against one's obviously heartfelt testimony, for the past is infinitely elastic.

In our present, imperfectly postmodern world, where most information still takes the potentially embarrassing form of printed matter lurking in archives, liars still must position themselves so that the historical record may not easily gainsay them. In that regard, UFOs have the advantage of goblins and ghosts, entities known to be capricious, elusive, unverifiable in their very nature, whose existence is strictly a function of our willingness to credit the testimony of those who choose to tell such tales.

There are two regions that liars head for by preference: periods of convenient isolation and the remote past. Strieber was abducted from the bedroom of a rustic cabin in the Catskills, and other abductees have usually been similarly circumstanced. Another, grander kind of liar rewrites history on a cosmic scale, telling lies not about himself but about the entire planet from literally the day of creation. The great-granddaddy of such liars was Ignatius Donnelly (1831–1901), a man whose once-flourishing fame has withered to the size of a few footnotes in out-of-the-way scholarly texts. Donnelly wrote three SF novels, one of which, *Caesar's Column* (1889), was a best-seller in its day (and will be considered in Chapter 9, as a prototype of the *Star Wars* techno-thriller), but his true talent, his genius, was for hoaxing. He imposed on the credulity of the public on three separate occasions, and all three inventions, in mutated form, are still in circulation.

His first and most imitated fabrication was a work of pseudo-archaeology, *Atlantis: The Antediluvian World* (1882), in which he argued "that the description of this island given by Plato is not, as has been long supposed, fable, but veritable history," that it was "the region where man first rose from a state of barbarism to civilization, . . . from whose overflowings the shores of the Gulf of Mexico, the Mississippi River, the Amazon, the Pacific coast of South America, the Mediterranean, the west coast of Europe and Africa, the Baltic, the Black Sea, and the Caspian were populated by civlized nations."[15] In short, all recorded history is in error, except for Plato and the the Book of Genesis. (Even in 1882, Donnelly knew that the best way to pitch a flaky theory is to connect it with a tenet of fundamentalist faith. If you can believe in Noah's ark, why not Atlantis?)

Already in the nineteenth century, for a hoax to succeed, there had to be some semblance of "science" in the mix, and Donnelly cited evidence from the then infant science of archaeology: "Among the Romans, the Chinese, the Abyssinians, and the Indians of Canada the singular custom prevails of lifting the bride over the door-step of her husband's home." How to account for this? The only explanation must be these cultures' common source in the customs of Atlantis. For linguistic evidence there's this: "How can we, without Atlantis, explain the presence of the Basques in Europe, who have no lingual affinities with any other race on the continent of Europe, but whose language *is similar to the languages of America?*" The book is one great piñata of such specious correspondences between the alphabets, mythologies, folkways, and architectural artifacts of all civilizations. Whatever had glintingly caught Donnelly's magpie attention became another proof of our Atlantean origins.

From Donnelly's Atlantis has sprung a vast progeny of SF-flavored pseudohistories, the most popular of which has been Erich von Daniken's *Chariots of the Gods?* (1968). Von Daniken would have it that

> dim, as yet undefinable ages ago an unknown spaceship discovered our planet. The crew of the spaceship soon found out that the earth had all the prerequisites for intelligent life to develop. . . . The spacemen artificially fertilized some female members of this species, put them into a deep sleep, so ancient legends say, and departed. Thousands of years later the space travelers returned and found scattered specimens of the genus *homo sapiens.* They repeated their breeding experiment several times until finally they produced a creature intelligent enough to have the rules of society imparted to it. The people of that age were still barbaric. Because there was a danger that they might retrogress and mate with animals again, the space travelers destroyed the unsuccessful specimens or took them with them to settle them on other continents. The first communities and the first skills came into being; rock faces and cave walls were painted, pottery was discovered, and the first attempts at architecture were made.[16]

With this unsavory amalgam of Darwin, the Old Testament, and the eugenic fantasies of the Third Reich, von Daniken scored a huge publishing success. "Over 4,000,000 copies in print" brags the cover of the thirty-fifth paperback printing from 1978. Although Donnelly lived before the age of UFO mythology and paperbacks, there is nothing in *Chariots* that cannot be found already fully developed in *Atlantis* and in Donnelly's successor hoax, *Ragnarok: The Age of Fire and Gravel* (1883), which explains how a long-ago comet had almost collided with the

Earth, sinking Atlantis and wreaking assorted other havocs. This rather modest astronomical fantasy, which does for Newton what Donnelly had already done for Darwin, prefigures the work of Immanuel Velikovsky (e.g., *Worlds in Collision*), another redneck archaeologist who also offers the litter of ancient civilizations as proof of his ditzy theory that the solar system is like a game of croquet played by vengeful gods.

No doubt many of the readers of Strieber, von Daniken, and Velikovsky approach their books in the same playful spirit they would bring to an SF story, asking only to be amused. Their books offer larger servings of the campy pleasures available in supermarket tabloids that show photos of Clinton shaking hands with an alien. For such readers, "Far out!" "Weird!" and "What next?" are expressions of appreciation, and belief is not really at issue. Even most of the great mass of those who tell pollsters they believe in UFOs can best be understood to be "entertaining" that belief, partly because aliens are a nifty idea, as long as they never directly impinge on one's life, and partly because to profess such belief has become a way of giving the finger to know-it-all intellectual snobs.

A certain class of reader values bizarre and paranoid theories precisely because they are bizarre and paranoid. In the lates '70s the SF writer Robert Anton Wilson brought out a series of books under the umbrella title of *Illuminatus!* that aspired to be a Summa of all conspiracy, occult, and UFO theories. Some of the books were offered as fiction, some as nonfiction. For Wilson and his fans, veridity was never an issue. I saw him once, after a book signing in Los Angeles, gravely romancing a would-be true believer, throwing out dark hints, then lapsing into winks and giggles. Did he experience cognitive dissonance? I wondered at the time. Does Oliver Stone when he films egregious distortions of the historical record *as though* he were recreating actual events? To both questions the answer is: probably not. They must see themselves not as liars, or even romancers, but as poets, in the sense that Sir Philip Sidney intended when he wrote in his "Defense of Poetry" of 1595, "Only the poet . . . lifted up with the vigor of his own invention, doth grow, in effect, into another nature, in making things either better than nature bringeth forth, or, quite anew, forms such as never were."

The license that "poets" assume in rewriting ancient history to suit their own fancy and sense of cosmic justice is not always without unhappy consequences in the real world. Witness the effect that such fabulation has had on school and university programs throughout the country, where "African-American Baseline Essays" has been used as a text to teach students that ancient Egyptians (who were black) developed the theory of evolution long before Darwin, understood quantum mechanics, flew gliders, could predict auspicious days by astrology, and could foresee the future by their psychic powers. This information is passed off as science. Martin Bernal, the author of *Black Athena* (1987), would have us believe that Greek civilization was either borrowed or stolen from Egypt. Other Afrocentrists claim that Aristotle stole his philosophy from books in the Library at Alexandria (a city that did not exist in his lifetime); that Socrates and Cleopatra were black (a fact of which their many detractors made no mention). Commenting on these matters in the *New York Review of Books*, Jasper Griffin says: "These assertions and the persistence with which they are made in the face of refutation form a fascinating study in morbid collective psychology. . . . But the implications are worrying. Some academics now say, and others think, that it does not matter whether these assertions are based on evidence or not, or whether they do or do not stand up to dispassionate scrutiny."[17]

To put it another way, Afrocentric mythologizers have the right to lie. Not only that but to controvert or ridicule their spurious scholarship is an act of racism. Ten years ago it was Tawana Brawley's self-serving charges of being raped that were at issue; now it is Western civilization *tout court*. Those who are inclined to shrug must suppose that no one is harmed by such fantasies, which may serve, after all, as a valuable source of self-esteem for black students. Real harm is done by such charlatanry, however. Those bamboozled into believing palpable untruths that are recognized as such by the larger community are likely in time to develop an attitude of truculent resentment and outright paranoia rather than self-esteem. James Wolcott, reviewing a recent tome of UFO lore in the *New Yorker*, describes his own close encounters with "abductees":

> They bugged me. I came to feel that I was dealing with a quasi-cult of deluded cranks. The abductees I interviewed, far from being people

> plucked out of the ordinary workday, had browsed the entire New Age boutique of reincarnation, channeling, auras, and healing crystals. . . . For them the aliens were agents of spiritual growth [but beneath that] was a pinched righteousness; the ones I met tended to be classic pills of passive aggression. Anger and distrust brooded beneath the surface.[18]

Not all the aggression of UFO believers can be counted on to be passive, however. On June 14, 1996, three men on Long Island—one the president of the Long Island UFO Network, and the other two members of the organization—were arrested for plotting to assassinate Suffolk County officials and seize control of county government. What had spurred them to this act was the refusal of those officials to recognize the clear and present danger posed by the UFOs that had set brush fires on Long Island the previous summer.

The potential for such fanaticism is always present when people insist that their self-delusions, dreams, and lies must be taken at face value by the world at large. The world, alas, often refuses. The gentlemen on Long Island certainly made the wrong move in trying to resolve the tension inherent in that situation. Had they been wiser, and a little more patient, they would have done what other gifted liars have done, sometimes with wonderful success: they would have started their own religion.

Or, if they lacked that degree of grandiosity, they could have followed the career path blazed for them by American's first SF writer and one of our most accomplished liars, Edgar Allan Poe.

Chapter 2

POE, OUR EMBARRASSING ANCESTOR

Poe is the source.

Many others have been claimed as SF's essential ancestor, beginning with the nameless author of the legend of Gilgamesh, the first Superman. As well make the same claim for J, who wrote of Noah and the Flood, of Nimrod and the Tower of Babel—themes that will appear in many SF updates.[1] But if myths and legends are to be credited as SF, then half of world literature before the novel must be accounted ancestral to SF. Homer's description of an invisible Ulysses' entrance into the court of Alcinous prefigures a host of other invisible men. Lucian of Samasota described voyages to the moon and sun, and provided inspiration to such other proto-SF writers as Rabelais, Cyrano de Bergerac, and Jonathan Swift, all of whom recounted similar fantastic voyages. For those SF writers who write in a vein of off-the-wall whimsey or gross hyperbole—writers as diverse as Piers Anthony, R. A. Lafferty, and Robert Sheckley—Lucian and his lot may be considered spiritual forefathers but not, in a generative sense, lineal. Lucian was spoofing the Greek romance novels of his time and would have snickered to think of someone taking the idea of flying to the moon as anything but silliness. In much the same way, SF's satirists and farceurs have taken their impetus from pulp science fiction, using its stock figures—astronauts and aliens—for their own japes and gambols. Any resemblance to Rabelais is purely coincidental.

If one eliminates all claimants whose claim is simply a penchant for giving their fantasies the sheen of the verisimilar, there remains one significant rival to Edgar Allan Poe as the genre's founding genius: Mary Shelley, the author of *Frankenstein* (1818). Brian Aldiss, in his authoritative history of the genre, *Billion Year Spree*,[2] presents the best case that can be made for Shelley, citing her debts to Milton, Goethe, and Erasmus Darwin, by way of showing that she is not just another gothic novelist: "We can see," Aldiss urges [p. 26], "that Erasmus Darwin thus stands as father figure over the first real science fiction novel. The Faustian theme is brought dramatically up to date, with science replacing supernatural machinery. . . . Frankenstein's is *the* modern theme, touching not only science but man's dual nature, whose inherited ape curiosity has brought him both success and misery."

For all Aldiss's insistence on the significance of Shelley's themes, his arguments are as merely theoretical as those urged on behalf of Lucian. In the sense that Lucian imagined the moon as some*where* one might be sent, he wrote SF. Mary Shelley's notion, that human life could be engineered by a capable scientist, is SF in the same rudimentary way. But once that premise has been established with a bit of fast talking and a few stage props, Shelley's tale becomes a Model-T melodrama with philosophic interludes. The few memorable high moments are surrounded by great expanses of narrative tundra. Aldiss himself adduces the most compelling reason for denying Shelley full honors as SF's progenetrix when he observes: "For a thousand people familiar with the story of Victor creating his monster from selected cadaver spare parts and endowing them with new life, only to shrink back in horror from his own creation, not one will have read Mary Shelley's original novel." This suggests, to Aldiss, "the power of infiltration of this first great myth of the industrial age." It suggests to me the failure of Mary Shelley to do justice to her theme. An unread author is no one's intellectual ancestor.

Mary Shelley had the good luck to have parents who were intellectual savants and to have married a great poet who was both well-born and well-heeled. Without those advantages it's doubtful that *Frankenstein* would have stayed in print to become one of the curiosities of the Romantic movement. That is not to say that the book is without all merit, only that its appeal will always be limited to those who read fiction in a spirit of academic curiosity, not for entertainment, and the intellectual significance that Brian Aldiss ascribes to it will not have an impact on the experience of most readers. For the great majority, who know it only in its simplified and distorted film versions, *Frankenstein* is science fictional only in the reductive and pejorative sense that it concerns a monster who runs amok and poses a danger to defenseless women. He is Dracula without the glamour, a zombie energized by lightning, a figure in a costume almost two hundred years old pursued by a mob of villagers brandishing torches. Such imagery belongs not to science fiction but to its generic opposite, ancient history. The figure of the robot run amok or morosely pondering the hole where its soul should be, a figure made memorable in Arthur Clarke's *2001* and Isaac Asimov's *I, Robot* stories, descends not from Shelley but from the much later Karel Capek of *R.U.R.* fame, with perhaps a tip of the hat to Frank Baum's Tin Man of Oz.

Poe is the source, because people read his stories. Even in the decades just after his death in 1849, when he was grossly slandered by his literary executor, Rev. Rufus Griswold, in a memoir appended to the first collected edition of his work, Poe did not want for readers. Within a decade, the three-volume edition containing Griswold's hatchet job had reached its seventeenth edition, despite its author's being limned as a man who "exhibits scarcely any virtue in either his life or his writings. Probably there is not another instance in the literature of our language in which so much has been accomplished without a recognition or a manifestation of conscience."[3] Griswold's mud stuck, in part because Poe's life and character *were* assailable, but also because the Poe of Griswold's myth was the Poe his readers wanted: a starving artist, a devious liar, a drunkard, a sexual reprobate, and an all-around mad genius whom one could easily imagine inhabiting his own sensational fictions.

In Europe, Poe was applauded for the very qualities that inspired Griswold's opprobrium. Baudelaire idolized Poe as the perfected master of his own self-destruction, and generations of French writers took their lead from him. Nietzsche, Rilke, and Kafka were among his Germanic acolytes and professed similar reasons for their devotion. Europeans continue to look to America for a particular kind of literary reprobate to lionize. William Burroughs and Charles Bukowski, though not without honor in their own country, inspire a reverence on the Continent that is accorded only to those self-mythologizing writers who act out the stories they write. In their case, as in Poe's, the myth is that of the tenderfoot turned desperado; the bourgeois gentleman, well spoken and once well dressed, whose face is now on the barroom floor; such a tragic figure, the European reader might imagine, as he would become himself if he were to cross the ocean and revel in the honky-tonks of the Wild West.

American readers, those at least of an upper-middle-brow, are less likely to buy into the same myth. When the gutter is visible just outside one's own door, it's harder to romanticize. And so, though Poe was read by his own countrymen, he was read grudgingly. Contemporary taste-makers dismissed him out of hand. To Emerson he was "the jingle man." Whitman[4] allowed as how there was an "indescribable magnetism about the poet's life and reminiscences, as well as the poems," even admitted he was "brilliant and dazzling, but with no heat," but he finally withheld full honors on the ground (precisely echoing Griswold) that his work was "almost without the first sign of moral principle, or of the concrete or its heroisms, or the simpler affections of the human heart." While later critics were to be more generous than that, there was always an element of dismissal in even their highest praise. Poe's work was considered most suitable for intellectually hyperkinetic teenagers. T. S. Eliot put it this way:

> That Poe had a powerful intellect is undeniable: but it seems to me the intellect of a highly gifted young person before puberty. The forms which his lively curiosity takes are those in which a pre-adolescent mentality delights: wonders of nature and of mechanics and of the supernatural, cryptograms and cyphers, puzzles and labyrinths, mechanical chess-players and wild flights of speculation. The variety and ardour of his curiosity delight and dazzle; yet in the end the eccentricity and lack of coherence of his interests tire.[5]

What Eliot says of Poe can be said, in much the same terms, of science fiction as a genre. The golden age of science fiction is twelve, which is about the same age one is likely to discover Poe and dote upon him. I remember encountering "The Gold Bug" in sixth grade and developing, as per Eliot, a passion for cryptograms. Indeed, my first published work, for which I was paid $1.00 by a crossword magazine, was a letter-substitution encoding of some lines from Horace that I found in Bartlett's *Familiar Quotations*. By age twelve I'd read all of Poe's more popular tales and loved to terrify my younger brothers with my rendition of "The Tell-Tale Heart," which I read aloud to my siblings by the light of an upward-tilted flashlight so as to lend my face suitable ghastly shadows. Eventually, in high school, I committed the story to memory, along with many of Poe's poems. In the same years, and in much the same spirit, I became a fan of science fiction.

And, as well, of Shakespeare, Dickens, Ibsen, Thomas Hardy; of Beethoven, Wagner, Gershwin, Ravel, and Bach; of Picasso, Dufy, Rembrandt, and da Vinci: of Hitchcock, de Mille, Fellini, and Bergman. T. S. Eliot might have remarked in each of them, as in Poe, "the variety and ardour of his curiosity" as well as "an eccentricity and lack of coherence of his interests." No doubt Eliot is right in noting that a "gifted young person before puberty" delights in "wonders of nature and of mechanics." Those who beat the drum most loudly on behalf of science fiction often speak of the "sense of wonder" as the genre's specific virtue. But is it a mark of maturity and wisdom to lose that susceptibility? May not "the eccentricity and lack of cohesion of interests" that Eliot finds tiresome in Poe be a symptom rather of Eliot's narrowness than of Poe's too voracious appetite?

The most passionate (and prejudiced) critics tend to be those, like Eliot (and Poe), impelled to invent an aesthetic that validates their own work. What is remarkable is that Eliot found it necessary to denigrate Poe's work and not simply to ignore it. The high estimate of Baudelaire and other continental writers of similar status surely provided some impetus, but a more fundamental reason may be that Poe was Eliot's shadow in a Jungian sense. He represents a vein of modernism that is antithetical to that of Eliot and his compeers: one addressed not to a highly literate audience of university dons but to a much wider readership; one that reveled in excess, in downright bad taste just for the fun of it; one that adapted its art to the conditions of the marketplace without qualms, candidly motivated (as the poor commonly are) by a desire to earn money; and, finally, a modernism more interested in the future than the past, as one might expect of art tailored to appeal to a nation of immigrants.

The contest—Poe as against Eliot; science fiction as against "mainstream" literary fiction—is between two kinds of art, lowbrow and highbrow, popular and . . . What is the opposite? Unpopular? That would account for the need of Eliot and his kind to be alarmed by the barbarians at the gate. Writers wish, above all else, to be read, and so some of them must assist at the work of active denigration. Or is "elitist" a truer opposite? It is every populist's epithet of preference, and it's often quite true. If elitism means that one prefers ballet to hip-hop, Wallace Stevens to Rod McKuen, or Saul Bellow to a *Star Trek* novelization, then a case can be made for elitism—even without consigning all that is demotic or popular to perdition.

For the fact of the matter, which has become, a century and a half after Poe, *almost* conventional wisdom, is that highbrow and lowbrow are differences not of essential merit but of demographics. Eliot's modernism has an unassailable importance—but so does Poe's. Eliot has wholly unworthy epigones—and so does Poe. However (and this is the however that apostles of High Culture, such as Eliot, so bitterly resent), popular culture has more clout. Steven King, a lineal descendant of Poe, dominates the best-seller lists, and even *Star Trek* novelizations regularly outsell works by authors of high repute.

The difference between highbrow and low—between Eliot and Poe, between mainstream and sci fi—is not one that can be mapped by the conventional criteria of criticism. One cannot maintain, for instance, that Eliot is a formalist and Poe is not, for all writing is inherently formalist—Poe's more obviously so than that of Eliot. Which is the more "formalist" poem, "The Raven" or "The Wasteland"? In the matter of overt didacticism and preachiness, Poe is once again the more poe-faced. Eliot preaches at us quite as much as Emerson or any other high-minded Victorian (excepting Poe). Eliot sneers, in the passage quoted above, at the juvenile character of Poe's themes, but that sneer doesn't take into account those realms of Strangeness that mesmerized the likes of Baudelaire, Dostoyevsky, and Rilke.

The essential difference is not one of aesthetics or of some subtler metaphysical nature, but of the two writers' antithetical social and economic positions. Poe's writing was market driven. He was one of the first writers to aim at the mass market. It was his great misfortune that that market was in its infancy when he wrote, in the 1830s and 1840s. Poe's lifelong, and forlorn, hope was to create a magazine of nationwide scope that would cater to a modern readership with a short attention span (he believed that stories and poems should be read at one sitting) and an appetite for lurid sensation. He continually wrote and rewrote prospectuses for this ideal magazine and tried (and failed) to find financial backers.

Poe was a magazinist. The *Oxford English Dictionary* dates the word's first appearance to 1821, and it has probably always been a pejorative—one who writes for magazines. Now we would simply say a hack. A magazinist is significantly different from a novelist. Novelists are so called because they write novels. Magazinists do not write magazines; they write *for* magazines. They write to meet the demands of an existing audience. It was Poe's unhappy fate to be writing for his audience before his audience knew that it was there, and he died in penury as a result. But he was, at least prophetically, right. The audience existed, and it would come, over the course of time, to have dominion over the written word. In a letter trying to drum up funding for his ideal magazine, Poe declared:

> Holding steadily in view my ultimate purpose—to found a Magazine of my own, or in which I might have a proprietary right, it has been my constant endeavor in the meantime not so much to establish a reputation great in itself as one of that particular character which should best further my special objects. Thus I have written no books and have been so far essentially a Magazinist.[6]

Poe's ideal magazine, to judge by his many prospectuses and his own practice as editor and writer, would have combined the best qualities of *The New Yorker* with the worst of *National Enquirer*—which is to say that Poe had nothing against art so long as it was sufficiently flashy, but that he was also ingenious in catering to what he perceived to be his audience's need to be informed that its hopes and fears (along with its silliest daydreams and its most paranoid suspicions) were well founded. And it is in this particular regard that Poe can be considered the true source of modern science fiction.

It will be the task of most of the rest of this book to show in what respect the science fiction of the last half-century conforms to this template of prophesying to the converted and assisting in the deceptions of the self-deceived. But let us see first how the genre's great-grandfather went about that task.

Poe was a hoaxer. His first and most successful hoax (and his first work of proto-SF) was an 1831 newspaper story in the *New York Sun* purporting to be the journal of a Welsh balloonist who is swept from his intended course to Paris and fortuitously accomplishes the first transatlantic flight. Poe begins by mimicking, presciently, the tone of agitated solemnity of a *March of Time* documentary of the 1940s:

> The great problem is at length solved! The air, as well as the earth and the ocean, has been subdued by science, and will become a common and convenient highway for mankind. *The Atlantic has been actually crossed in a Balloon!* and this too without difficulty—without any great apparent danger—with thorough control of the machine—and in the inconceivably brief period of seventy-five hours from shore to shore!

Poe then skillfully segues from breathless to "bare facts":

> By the energy of an agent at Charleston, S.C., we are enabled to be the first to furnish the public with a detailed account of this most extraordinary voyage, which was performed between Saturday, the 6th instant, at 11 A.M. and 2 P.M., on the Tuesday the 9th instant, by Sir Everard Bringhurst; Mr. Osborne, a nephew of Lord Bentinck's; Mr. Monck Mason and Mr. Robert Holland, the well-known aeronauts; Mr. Harrison Ainsworth, author of "Jack Shephard," etc; and Mr. Henson, the proprietor of the late unsuccessful flying machine—with two seamen from Woolwich—in all eight persons.

Imagine yourself in 1831, picking up a copy of the *New York Sun*: wouldn't you believe it? Just as we (we SF readers) all knew, well before *Sputnik*, that space flight was inevitable, so the *Sun*'s readers knew that someday the Atlantic would be spanned by a manned airship, as the English Channel had been already. It was only a matter of time, and the time had come. Apparently.

Poe's hoax succeeded—at least until a confirmation from South Carolina was not forthcoming. According to Daniel Hoffman, the author of the best book-length critical study of Poe, *Poe, Poe, Poe, Poe, Poe, Poe, Poe,* the span of time between publication and disclaimer was probably the happiest day of Poe's life. "Hoaxiepoe," Hoffman dubs him, and declares that Poe's greatest impostures were not those that dealt with imaginary voyages into maelstroms or across oceans but "scientific ghost tales" that represent the conquest of time:

> Although now nearly forgotten, these [sketches] were what first attracted Baudelaire to Poe—the French poet's first translation of Poe's work was a rendering of "Mesmeric Revelations," a tale on Swedenborgian journal welcomed as a genuine report of a scientific experiment.
>
> Ah, those innocents. So great was *their* need to believe anything which purported to prove the existence of the soul that they were taken in by Hoaxiepoe. Like the editors and readers of the New York *Sun* in the matter of the balloon-crossing of the Atlantic. But, admits Poe, "The story is a pure fiction from beginning to end."[7]

Poe wrote "Mesmeric Revelation" and its companion tale, "The Facts in the Case of M. Valdemar," in 1844 and 1845, respectively. Though only in his mid-thirties, most of his best work was already behind him. His once child-bride, Virginia, whom he'd married when she was thirteen, was dying of tuberculosis, the disease that had killed both his mother and brother. The fear of death—one's own and that of those one loves—had been Poe's constant theme in his fiction and poetry. He was not alone, of course. The Victorians obsessed about death. The protocols and rituals of mourning were highly codified. Popular novelists like Dickens and Stowe prided themselves on their tear-jerking death scenes. Poe had his own vein of maudlin lamentation, which is most evident in his poetry, where he commemorates, with full funeral pomp, "the rare and radiant maiden whom the angels name Lenore." Yet even in this, his most celebrated poem, the Raven's incessant refrain of "Nevermore" would deny the poet's pious wish of his ultimate reunion with his sainted maiden "within the distant Aidenn." Death, Poe insists monotonously, is final, and the angels and Aidenn and even, alas, the maiden's sainthood are all just wishful thinking, a courtly fiction invented for the solace of genteel ladies.

Poe was surely not alone in that dismal view, though few other writers of the Victorian age were so candid in expressing it. The Enlightenment had happened, and it could not be made to unhappen. Doubts had been entertained, and while it was no longer considered good form to express those doubts after the manner of Voltaire and that lot, the Christian sentiments to be found in the leading writers of the time—in Tennyson, Emerson, Dickens, George Eliot, Hugo, Tolstoi—are chiefly just that: sentiments. Jesus is admired as a moral exemplar, but his divinity has become a polite fiction, much like the monarchy in England. The crux of the Christian faith was always its program for an afterlife of rewards and punishments, and *that* center was not holding.

The problem with the afterlife is that it remains unproved. It is the country, as Hamlet complains, from which no traveler returns. That, of course, never stopped poets from offering their own eyewitness accounts. From Dante Alighieri in the fourteenth century to the lately deceased James Merrill, poets have found that nothing pleases readers so much as a plausible and detailed account of the afterlife. Indeed, poetry isn't necessary; the bare assertion is enough.

"Mesmeric Revelation" and "The Facts in the Case of M. Valdemar" are such bare assertions—artfully artless, written in the language of popular science reporting. The appearance of scientific objectivity is crucial to such impostures, since it was science, after all, that first displaced faith by casting doubt on a literal subterranean hell and a celestial heaven, such as Dante had depicted. Poetic justice demands that science should assist at the resurrection of the afterlife, and Poe, with prophetic genius, found just the science to do the job: mesmerism.

In 1844 mesmerism, or, as we know it, hypnotism, had already been around for three-quarters of a century. Its discoverer, Franz Anton Mesmer, an Austrian occultist and proto-psychiatrist, had ascribed his successes in quieting hysterical patients to his channeling of the healing and magnetic power in his hands into the bodies of his patients. Accused of practicing magic, he emigrated to Paris, where he was denounced as a charlatan and investigated, in 1784, by a committee of doctors and scientists (including Benjamin Franklin), who gave him a thumbs-down. Mesmer was forced into retirement, but mesmerism continued to exert its fascination, and by the time Poe came to write "Mesmeric Revelation," he could begin his tale with this confident assertion:

> Whatever doubt may still envelop the *rationale* of mesmerism, its startling *facts* are now almost universally admitted. Of these latter, those who doubt, are your mere doubters by profession—an unprofitable and disreputable tribe. There can be no more absolute waste of time than the attempt to *prove*, at the present day, that man, by mere exercise of will, can so impress his fellow, as to cast him into an abnormal condition, in which the phenomena resemble very closely those of *death*.

The tale goes on to recount how one Mr. Vankirk, on the brink of death from tuberculosis, summons the author to his deathbed so that he may be interrogated while in a mesmeric trance. Vankirk explains:

> "I sent for you tonight . . . not so much to administer to my bodily ailment, as to satisfy me concerning certain psychal impressions which, of late, have occasioned me much anxiety and surprise. I need not tell you how sceptical I have hitherto been on the topic of the soul's immortality. I cannot deny that there have always existed, as if in that very soul which I have been denying, a vague half-sentiment of its own existence. But this half sentiment at no time amounted to conviction."

Vankirk believes that if he is interrogated while in a mesmeric trance, his doubts in the matter of his own immortality may be resolved. The author accedes to his request, and "a few passes threw Mr. Vankirk into the mesmeric sleep," from which vantage he explicates—sometimes with an artful inarticulacy; at others, with penny-a-word prolixity—Poe's own pseudoscientific ruminations about the Other Side.

For all his persiflage about "unparticled matter" and "the luminiferous ether," where spirit and matter become congenially blurred (much as they do nowadays in those books that explain how quantum physics and tai chi are all holistically One), Poe will persuade only those longing to be buffaloed. Poe explains that matter is like spirit, but different; that our senses see through a glass darkly, and that when he says the mesmeric state resembles death, "I mean that it resembles the ultimate life; for when I am entranced . . . I perceive external things directly, without organs." How's that again? you may ask. Poe's mesmerized, ventriloquistic dummy answers:

> The organs of man are adapted to his rudimental condition, and to that only; his ultimate condition, being unorganized, is of unlimited comprehension in all points but one—the nature of the volition of God—that is to say, the motion of unparticled matter. You will have a distinct idea of the ultimate body by conceiving of it to be entire brain. This it is *not;* but a conception of this nature will bring you near a comprehension of what it *is.* A luminous body imparts vibrations to the luminiferous ether.

There, in that marvelous conflation of wish fulfillment, pop theology, and pseudoscientific persiflage is all science fiction in a nutshell—and, intrinsic to it, the better part (and the worst) of the American Pop Culture to come. It is, at once, a wonderful and prophetic achievement and a cheap shot.

Although I anticipate some of the arguments I mean to develop at length later, let me count the ways that Poe, in this single tale, anticipates the entire genre of SF.

1. *Mesmerism.* Unerringly, Poe has lighted on the intellectually fuzziest area of the then embryonic "science" of psychology. Hypnotic "testimony" is still, at the end of the twentieth century, the preferred modus operandi for those who would have us believe in psychic regression to our previous lives, in the ritual Satanic abuse of children, and in UFO abductions, for the naive presumption persists that one who has been hypnotized cannot invent or confabulate. It is a license for liars.

2. *Dreams come true.* The assurance that Poe offers in "Mesmeric Revelation" is an immemorial promise that has resisted proof through recorded time: there is no death. Immortality is one of the favored

themes of SF, but the genre is home to many others equally unprovable and no less seductive: psychic powers of various kinds—telepathy, telekinesis, and all the other ways of having wishes come true by wishing really hard.

3. *Chip-on-the-shoulder superiority.* SF is a lumpen-literature. That can be a strength, but it is usually the genre's Achilles' heel. Poe, in the tale's second sentence, anticipates his critics by saying, "Those who doubt are your mere doubters by profession—an unprofitable and disreputable tribe." So say legions of UFO abductees. So say, as well, other SF true believers of various faiths—those who believe we *must* terraform Mars, those who've slid from science fiction into Scientology, those who have psychic powers but, even so, low-paying jobs. Poe had the same problem.

4. *Genuine visionary power.* And there's the rub—and the reason that science fiction bears pondering as more than a passing relic of the Zeitgeist like hula hoops and baseball cards. Poe was on to something. Even in "Mesmeric Revelation" he has moments when he's more than Hoaxiepoe. That notion of thinking of "the ultimate body" as nothing but the "entire brain" has become an icon of the twentieth century. Read Roland Barthes' essay, "The Brain of Einstein," or price out the movie poster for "Donovan's Brain."[8] Or catch the British SF writer Charles Platt on a TV talk show seriously discussing his intention to have his head cut off when he's in the condition of Poe's Mr. Vankirk so that it may be quick-frozen and eventually revived to proclaim its own "cryogenic" revelation.

The most visionary moment in "Mesmeric Revelation," and the one that foreshadows SF's most essential image—the Alien—has gone virtually unnoticed, no doubt because of Vankirk's tendency, even under hypnosis, to be hyperpolysyllabic:

> The multitudinous comglomeration of rare matter into nebulae, planets, suns, and other bodies which are neither nebulae, suns, nor planets, is for the sole purpose of supplying *pabulum* for the idiosyncrasy of the organs of an infinity of rudimental beings. But for the necessity of the rudimental, prior to the ultimate life, there would have been no bodies such as these. Each of these is tenanted by a distinct variety of organic, rudimental, thinking creatures.

What Poe is saying is exactly what Carl Sagan and generations of other SF writers before him have been saying: the universe is so big that there *must* be other life out there, and so we are not alone. There must be a kind of teleology in the universe that makes life and consciousness inevitable.

This may seem like cold comfort to those of conventional religious belief, but for a post-Enlightment doubter, the notion that our human consciousness is a precedent for the same thing (life) happening elsewhere in a well-nigh infinite universe can be encouraging. Indeed, that idea has become an article of faith within the SF confraternity, and far beyond. Hundreds of millions of people around the world profess to believe in the reality of UFOs.

The significance of this, with regard to Poe, is that he understood that the older tenet of faith, an afterlife, is related in an essential way with this more novel concept, Alien Life; that one belief assists the other. The Hamlet who doubted his own afterlife also declares to Horatio that there are more things in heaven and earth than are dreamed of in his philosophy. Poe made it his particular mission (as does SF) to catalogue those other things Horatio doesn't dream of.

Visionary power is a dicey criterion in one deflating regard: it's utterly subjective. One man's vision is another's pipe dream. Indeed, in the Romantic century, poets were beginning to make the connection between visions and (opium) pipe dreams a regular part of their repertoire. Coleridge, De Quincey, Poe, and Baudelaire were all abusers of that drug and, ultimately, abused by it.[9]

5. *Great special effects.* Poe was a prophetic and exemplary popular artist in the sense that he refused to recognize the boundaries of good taste. He delighted in going over the top and grossing people out, and his readers delighted in this side of his work more than in any other. "Mesmeric Revelation" is too cerebral a tale to represent this side of his talent (and is, accordingly, not one of the public's favorites), but its companion tale, "The Facts in the Case of M. Valdemar," offers a good example of Poe at his gore-blimiest (a special form of the sublime). In this story, another victim of tuberculosis agrees to be mesmerized while on the brink of death; this time the result is not a theosophical colloquy but the zombification of the hypnotic subject in a state of suspended inanimation. The story's focus is on the symptoms of the "sleep-waker's" physical decay

during the weeks of his trance state: the rictus of the lips, "the mouth widely extended, and disclosing in full view the swollen and blackened tongue," "the profuse out-flowing of a yellowish ichor [from the eyelids] of a pungent and highly offensive odor," and a timbre of voice that impresses the narrator "as gelatinous or glutinous matters impress the sense of touch." At last the hapless M. Valdemar implores: "For God's sake!—quick!—quick!—put me to sleep—or, quick!—waken me!—quick!—*I say to you that I am dead!*" The end of the tale quickly follows:

> As I rapidly made the mesmeric passes, amid ejaculations of "dead! dead!" absolutely *bursting* from the tongue and not from the lips of the sufferer, his whole frame at once—within the space of a single minute, or even less, shrunk—crumbled—absolutely *rotted* away beneath my hands. Upon the bed, before that whole company, there lay a nearly liquid mass of loathesome—of detestable putridity.

A large part of the tale's first readers accepted it (as they had the "Mesmeric Revelation") as being the transcription of a real event. Even those not completely taken in declared, as did one Boston reader, "I have not the least doubt of the *possibility* of such a phenomenon." But the cognoscenti responded much as a fan of Hitchcock's might delectate over one of his more bravura passages. Virginia poet Philip Pendleton Cooke praised the story as

> the most damnable, vraisembable, horrible, hair-lifting, shocking, ingenious chapter of fiction that any brain ever conceived, or hand traced. That gelatinous, viscous sound of the man's voice! there was never such an idea before. . . .
>
> I have always found some remarkable thing in your stories to haunt me long after reading them. The *teeth* in Berenice—the changing eyes of Morella—that red & glaring crack in the House of Usher—the pores of the deck in the Ms Found in a Bottle—the visible drops falling into the goblet in Ligeia.[10]

In the early 1960s, Roger Corman directed a series of low-budget horror movies based on seven of Poe's most popular titles (including two of his poems, a unique distinction for any poet). Corman's aesthetic was exactly Poe's: lots of trumpery red velvet and fog machines all leading up to a payoff that is memorably gruesome, grotesque, or simply disgusting, according to one's taste in these matters. I particularly remember the hero of *The Premature Burial* lifting a chalice to his parched lips, only to discover that it is filled with writhing maggots.

Corman's Poe-inspired movies are representative of horror and science-fiction movies in general. Movie audiences want just what Philip Pendleton Cooke did: something "horrible, hair-lifting, shocking, ingenious." They want to see Godzilla-level Tokyo, and they want to *believe* it ("vraisembable," as Cooke has it). They want to see the alien's insectile claw *erupt* from inside a human stomach. They want to see exploding heads (as in Cronenberg's *Scanners*) and zombies in every stage of decay. They want to see Atlantis and Mars and the hollow core of the Earth. And anyone who can supply such needs well enough, in either prose or papier-mâché, is ensured a decent living in Hollywood.

6. *Sophomoric humor.* Poe's gross-out aesthetic is notably less successful in the realm of humor (to which he devoted a great part of his writerly energies) than in that of horror. Indeed, the gross-out is pretty much the only way Poe knows to be funny.

The aesthetic of the gross-out is often misunderstood. The gross-out artist does not aim to provoke laughter—or even a smile. The gasp is what he's after: the gasp of, *This isn't happening!* the gasp of one who witnesses what critics now like to call "transgressive" behavior, the indignant gasp of Margaret Dumont as she is goosed, once again, by Groucho. The comedians John Belushi and Andy Kaufman were gross-out artists of the first degree; so, too, in their own ways William Burroughs and the "performance artist" Karen Findley. And so, flamingly, was Poe.

Poe's most remarkable offerings in this vein are not among his best-known tales. Those who assemble anthologies and textbooks allow horror stories only when they are written in a tone of hushed reverence, not in one of madcap glee. Garrulous maniacs are permissible, like the narrators of "The Tell-Tale Heart" and a "A Cask of Amontillado," but not outright zanies, like the preposterous Signora Psyche Zenobia, who begins "A Predicament" in this wise:

> It was a quiet and still afternoon when I strolled forth in the goodly city of Edina. The confusion and bustle in the streets were terrible.

> Men were talking. Women were screaming. Children were choking. Pigs were whistling. Carts they rattled. Bulls they bellowed. Cows they lowed. Horses they neighed. Cats they caterwauled. Dogs they danced. *Danced!* Could it be possible? *Danced!* Alas, thought I, *my* dancing days are over!

Is not that truly god-awful? And in just the next, grotesquely obese sentence it gets worse, and continues to worsen to the very end in all the ways Poe can think to gross out a reader of refined sensibilities.

Psyche Zenobia is attended by a poodle and her servant, Pompey:

> Pompey, my negro! sweet Pompey! how shall I ever forget thee? I had taken Pompey's arm. He was three feet in height (I like to be particular) and about seventy, or perhaps eighty years of age. He had bowlegs and was corpulent. His mouth should not be called small, nor his ears short.

That too gets worse. Poe was an antebellum Virginian, and his view of blacks is quite simply that they were subhuman. In his fictions they figure either as objects of horror (the cannibals in his one novel, *The Narrative of* A. *Gordon Pym*) or, more commonly, as figures of fun.

Psyche is possessed by a desire to enjoy the view from the top of "a Gothic cathedral—vast venerable, and with a tall steeple." She mounts the interminable stairs and, attaining the belfry, bullies Pompey (by pulling out his hair) into letting her stand on his shoulders so that she might stick her head out an opening high on the wall. She rhapsodizes over the view until she is "startled by something very cold which pressed with a gentle pressure on the back of my neck." It is the minute hand of the steeple's clock, and Psyche describes, in the language of gothic horror, how the minute hand cuts through her neck so that first one and then another of her eyes pop from her head, which then is severed from its body, all of which is described with prissy exactitude. Pompey flees; Psyche's poodle is eaten by a rat. The tale ends with a moronic parody of German verse mourning the poodle's death, and then: "Sweet creature! she too has sacrificed herself in my behalf. Dogless, niggerless, headless, what now remains for the unhappy Signora Psyche Zenobia? Alas—*nothing!* I have done."

Were any undergraduate satirist to publish "A Predicament" as his own work in a campus magazine, it would be grounds for expulsion. That, in a way, is its glory. More than a century and a half after it was written, it still offends against all known proprieties except chastity.

Even so, Daniel Hoffman singles out this story as one of Poe's most telling works, partly for the way it lays bare the aesthetic of his more serious gothic tales (it was written as a supplement to his essay, "How to Write a Blackwood Article," that magazine being then the flagship of the expiring gothic tradition), and partly because its image of someone decapitated by the minute hand of a gigantic clock had invaded Hoffman's own nightmares as a teenager and had become a personal obsession.

That is the power of the successful gross-out: it is indelible. Think of Dan Aykroyd impersonating Julia Child and bleeding to death all over his/her haute cuisine. Or the Coneheads—those perfectly assimilated aliens—consuming mass quantities of potato chips and beer. Or Slim Pickens riding an H-bomb out of the opened hatch at the end of *Dr. Strangelove* and bidding the earth one last "Yahoo!"

It is no accident that the last two examples should be science fictional. The genre has always aspired to outrage genteel taste. Harlan Ellison's two resoundingly successful anthologies, *Dangerous Visions* (1967) and *Again, Dangerous Visions* (1972), owe much of their cachet to the editor's determination to publish stories too taboo ("transgressive" had not then been discovered) to be published anywhere else. But long before Ellison recognized the benefit of an X rating, SF writers had been champing at the bit. Philip Jose Farmer made his name by imagining the disgusting ways aliens might have sex, and Theodore Sturgeon was there, so to speak, at his side.

Not all of SF's systematic transgressions have been in a vein of gross-out humor—except insofar as the gasp at the moment of violation may lead to retrospective giggles. People often react to images of horror as, by his own account, Daniel Hoffman did. He'd read "A Predicament," dismissed it as silliness, forgot having read it, and then was haunted in nightmares by its central image until, years later, he rediscovered the story and was exorcised.

7. *Divine madness.* "True!—nervous—very, very dreadfully nervous I had been and am; but why will you say that I am mad?" So opens what

must be Poe's most widely known tale (unless that honor belongs to "The Fall of the House of Usher"), "The Tell-Tale Heart." Its narrator's paranoid grandiosity, as he confesses his misdeeds in a tone of bristling self-righteousness, has become a standby not only of pulp fiction and grade B movies but of the American legal system, where the most fantastic sophistries, such as the "Twinkie defense" of Harvey Milk's murderer, may be adduced to plead the innocence of those who have schemed to kill wives, parents, children, and pet peeves. The object of such a defense is not (I would theorize) to convince a jury that one is not guilty but rather, as Poe does, to create the spectacle of a guilt so beguilingly brazen that some jurors (one is all it takes) may react as Poe wants his readers to: by vicariously identifying with a glamourous or eloquent criminal and, in a spirit of perverse complicity, letting him slip free the snares of the law.

Four of Poe's most popular tales take the form of a madman's confession to the crime of murder: "The Black Cat," "The Tell-Tale Heart," "The Cask of Amontillado," and "The Imp of the Perverse." Other narrators are mad without being homicidal, men who act in a bizarre way while in the grip of obsessions or delusions, so that each of them might at some point echo the lines above: "Why will you say that I am mad?"

Poe is not alone in finding it good sport to feign madness. Many drunks (including Poe) drink so that they may behave like drunks. Opium and other drugs are used to provide recreational hallucinations. Mediums and channelers offer the credulous the solace of socially sanctioned delusions, or if one's budget is more limited there are pentecostal churches that encourage speaking in tongues. Various forms of psychotherapy encourage the simulation of madness in a controlled environment: primal screaming, hypnotic regression, and others. In all these cases, the "sane" person is able to take a holiday from sanity and enjoy freedoms of action or of the imagination denied by workaday social and moral codes.

Such sanctioned round-trip tickets to madness and back have become so commonplace that for a time, in the '60s and '70s, it was fashionable among radical psychotherapists, such as R. D. Laing, to maintain that madness was a higher form of wisdom, a kind of inadvertent shamanism. As a corollary, mental hospitals were represented in novels and movies of that period—most notably, Ken Kesey's *One Flew over the Cuckoo's Nest* (1964; film 1975), but also Peter Weiss's *Marat/Sade* (film 1966) and Philippe de Broca's *King of Hearts* (1966)—as lay monasteries where a few holy fools lived in a state of higher sanity, while in the world outside the asylum, the truly insane busied themselves with wars and revolutions.

One of Poe's "humorous" tales, "The System of Doctor Tarr and Professor Fether," has a good claim to being the source of these parables of asylums where the patients have taken charge—that being the premise of the story. It's not one of his best. After setting the stage by introducing an unsuspecting visitor into the transvalued asylum, Poe badly fumbles the ball. His madmen and madwomen, who are presented to the visitor as fellow guests at a stately dinner party, compete with each other at trumpeting their manias at the first opportunity. One woman insists upon telling how an earlier patient had "found upon mature deliberation, that, by some accident, she had been turned into a chicken-crow; but, as such, she behaved with propriety. She flapped her wings with prodigious effect—so-so-so—and, as for her crow, it was delicious! Cock-a-doodle-doo!—cock-a-doodle-doo!—cock-a-doodle-de-doo-doo-do oo-do-o-o-o-o-o-o!"

This is Poe in his broadest vein of gross-out humor, giving his readers what he supposes they want: a picture of lunatics misbehaving with gaga abandon. The story Poe *might* have written can be caught only in glimpses, as when the narrator sits down to listen to a beautiful young woman singing a Bellini aria and then has a polite conversation with her, wondering all the time whether she's sane:

> In fact, there was a certain restless brilliancy about her eyes which half led me to imagine she was not [sane]. I confined my remarks, therefore, to general topics, and to such as I thought would not be displeasing or exciting even to a lunatic. She replied in a perfectly rational manner to all that I said; and even her original observations were marked with the soundest good sense; but a long acquaintance with the metaphysics of *mania* had taught me to put no faith in such evidence of sanity, and I continued to practice, throughout the interview, the caution with which I commenced it.

Such a sly commentary on the tightrope we walk in our most ordinary "polite" conversations is worth a thousand cock-a-doodle-doos.

The "metaphysics of mania" has been, quite as much as the conquest of space, a staple of science fiction, often in the form of a story where the mutant hero is regarded as a madman, whereas (in the words of "The Tell-Tale Heart"), "The disease had sharpened [their] senses—not destroyed—not dulled them." Olaf Stapleton's *Odd John* is such a psychic superman, and Robert Silverberg's all-too-vulnerable telepath in *Dying Inside* another. Closer in spirit to Poe are tales of paranoid delusions that come true, such as Robert Heinlein's stories "They" and "The Unpleasant Profession of Jonathan Hoag." What if, Heinlein's stories ask, no one but oneself really existed? What if the whole universe were just a programmed delusion one has been tricked into believing?

Many readers find such solipsist speculation comforting rather than threatening. A lapel button popular in SF's fandom in the '60s declared: REALITY IS A CRUTCH. Those buttons may have become collectibles, but surely the same sentiment is operating again in today's vogue for the nascent technologies of virtual reality, a concept that has spawned dozens of SF novels on the same themes of Do I Wake or Sleep?/ Is It Real or Is It Memorex? Phil Dick was SF's all-time grand master of this subgenre of pop epistemology, and we will be dealing with him in Chapter 4.

Meanwhile, it should be pointed out that the pleasures of self-delusion are by no means limited to SF. They are intrinsic to fiction as such, and perhaps to all the arts insofar as they are mimetic. What sculptor of the female form is not, at heart, another Pygmalion? Even that most rock-solid and utilitarian of the arts, architecture, counts as its highest achievements the creation of sacred precincts that are meant to be heavens cast in marble or concrete, and the most popular (and specifically American) of all tourist attractions are theme parks after the model of Disneyland—Potemkin villages designed to deceive the paying customers. Surely it is no coincidence that three of the most notable successes of America's best-selling SF writer—*Westworld, Jurassic Park*, and *The Lost World*—have theme parks as their setting.

To speak of an art form as "popular" is another way of saying it is commercial. Perhaps, as Dr. Johnson declared, all writers, if they are not fools, write for money. But there is a great difference, in Poe's time and ours, between "commercial fiction" and the art of the novel as practiced by the better sort of novelist. Essentially it is a class distinction. Commercial fiction panders to low tastes and traffics in scandal, violence, and sentimentality; the art novel appeals to cultivated tastes and traffics in the same commodities, but in a more genteel way.

That Poe was poor, wretchedly poor, a starveling even at the height of his fame, is essential not only to his myth—few other artists of the first rank have been his equal in penury—but to the work he produced. He was the progeny of itinerant actors: a father who absconded, a mother who died soon after of TB (as would his wife and elder brother). He was adopted by a Virginia merchant, educated among the gentry, and attended West Point briefly. Then, having too often disgraced himself, he resumed the station in life he was born to, becoming an artist, a pauper, and a poseur. Think of him as the cousin of Twain's duo of pioneer performance artists in *Huckleberry Finn*, the Duke and the Dauphin, with the same theatrical airs, the same outsider resentments, the same foxy delight in taking rubes for a ride.

SF writers have been, by and large, Poe's heirs in these respects as well. They cater to an audience that demands to be diddled: the pilgrims to Roswell, New Mexico, that Oz of UFO believers; the buffs of Atlantis and Mu; the followers of L. Ron Hubbard; the Heaven's Gate and Aum Shinrikyo cults. All these cults owe their origins, more or less directly, to the specific fabulations of SF writers. Even the Roswell case, which would seem a genuine instance of "mistaken identity," has its component of science-fictional fraud. Robert Spencer Carr (the brother of celebrated mystery writer John Dickson Carr) began his writing career as a child prodigy, penning a best-selling novel at age seventeen; he produced SF hackwork through the early '50s, sank from sight, and became famous, briefly, in the '70s when, in a radio interview, he concocted the still-current story of aliens' autopsied and kept in cold storage at the Wright-Patterson Air Force Base, near Dayton, Ohio. Carr, on the testimony offered by his son Timothy, was an inveterate romancer: "Often he mortified my mother and me by spinning preposterous stories in front of strangers . . . [tales of] befriending a giant alligator in the Florida swamps, and sharing complex philosophical ideas with porpoises in the

Gulf of Mexico. It wasn't the tall tales themselves that hurt so much but his ferocious insistence that they were true. . . . They were dead serious, and you had by God better pretend you believed them or face wrath or rejection."[11]

Poe is the source not only of science fiction, but of the first nationwide hoax to have its origins in SF and one that has continued to this day. Spiritualism was the UFOlogy of the nineteenth century, and Poe provided much of its rationale in his "Mesmeric Revelation" of 1844. In 1849, a year after Poe's death, two teenage sisters in upstate New York, Maggie and Kate Fox, turned Poe's theory into practice. The Fox sisters astonished their family, neighbors, and eventually the nation by a talent for being haunted by a ghost that communicated through rappings—as of someone gently tapping, tapping at their chamber door. These spectral sounds (which the girls produced by the cracking of their toe joints against the floor) soon evolved into a crude kind of Morse code, and the Fox sisters became the world's first mediums, offering the bereaved the consolation of direct communications with the dead. As per Poe's "Mesmeric Revelation," the afterlife revealed by the hundreds of mediums who soon discovered that they shared the Foxes' psychic powers was a kinder and gentler place than the hells hinted at by earlier generations of revenant spirits. Mediums were not in the business of frightening away those who attended their seances. Their customers wanted comfort, and that's what they got, with just a frisson of spookiness.

As with UFOs, the stakes kept getting higher, and close encounters had to be arranged with the playing of spectral music, movements of the furniture, and fleeting visions of ectoplasmic spirits. The history of spiritualism is a constant repetition of the same pattern, as one medium after another produces some new marvel, thrives for a while, and at last overreaches and is exposed as a fraud.

The greatest medium of them all was Helena Petrovna Blavatsky (1831–1891), founder of the Theosophical Society, an institution created in 1875 and still extant, though no longer making the headlines and scandals that its early history was so rich in. "Madame" Blavatsky, as she is usually referred to, was the self-mythologizing author of *Isis Unveiled*

(1877), a "nonfiction" melange of the occult romances of Bulwer-Lytton: *Zanoni* (1842) and *A Strange Story* (1862). Peter Washington, the author of a vastly diverting group biography of Madame Blavatsky and her occult descendants, maintains, "It would not be unjust to say that her new religion was virtually manufactured from his pages."[12]

The other ingredient of *Isis Unveiled* and *The Secret Doctrine* (1888), which sets them apart from previous occult literature, was their challenge to the evolutionary theories of Charles Darwin, then as now the particular bogey of those who resented the intrusions of science on the realm of religious dogma. Rather than champion the revelations of the Bible, Madame Blavatsky wrote her own, an amalgam of Buddhism, Hinduism, and the Egyptian religion as she imagined it. Blavatsky tolerates some of Darwin's ideas, but reveals that evolution didn't end with the transformation of monkeys into men but is still at work transforming men into higher beings, like herself and her various spirit guides, the Great White Brotherhood of Masters. It was these higher beings—Serapis and Tuitit Bey, the Tibetan prince Master Morya, and the Kashmiri Brahmin Koot Hoomi (who had been Pythagoras in an earlier incarnation)—who dictated her books to her, or simply "precipitated" the completed manuscripts on her desk while she slept; they also consulted with her in public seances and left letters in the pockets of her disciples, urging them to exert themselves in providing her material needs.

She was, in short, a charlatan of indomitable chutzpah, and though, like other such, she was often caught in the act during seances, she dismissed such frauds as part of the Masters' larger plan, tests of loyalty and faith. "What is one to do when, in order to rule men, you must deceive them," she wrote, "when in order to catch them and make them pursue whatever it may be, it is necessary to promise and show them toys? Suppose my books and *The Theosophist* were a thousand times more interesting and serious, do you think I would have anywhere to live and any degree of success unless behind all this there stood 'phenomena'? I should have achieved absolutely nothing, and would long ago have pegged out from hunger."[13]

Brava! Spoken like a licensed liar! Not an *American* liar, it must be admitted, but she did learn her craft in America, when she journeyed to Vermont in 1874 to witness the performances of three child mediums who summoned a whole vaudeville show of apparitions: Indians squaws, recently deceased celebrities, dead children (whose mother attended every performance and was always thrown into amazement), and other spirits (seen in silhouette behind a curtain) who sang and danced and fought a duel with swords. Clearly the art of mediumship had come a long way since the Fox sisters began snapping their toes twenty-five years earlier.

Madame Blavatsky's heirs in the UFO era have recognized the same need to produce "phenomena" that will impress the guileless groundlings. Blavatsky offered levitations, spirit music, letters from the ether; UFO promoters simulate crop circles as evidence of saucer landings, create "alien autopsy" film footage, and script "reenactments" for tabloid TV. At the same time, by way of fudging the issue and attracting a higher class of clientele, a pseudoscientific rationale is deployed in combination with the incense of a New Age religion that is all effortless transcendence. The afterlife described by spirit mediums was a balmy and shadowless eden, with scarcely a whiff of brimstone. Such has increasingly become the tone taken even by professional UFO abductees. "I suspect," Whitley Strieber writes, "that the visitors may have been here for a long time. It has even crossed my mind, given their apparent interest in human genetics, that they may have had something to do with our evolution. . . . It is possible . . . that we are in the process of evolving past the level of superstition and confusion that has in the past blocked us from perceiving the visitors correctly."[14] His words might have come from the lips of Madame Blavatsky.

Chapter 3

FROM THE EARTH TO THE MOON—IN 101 YEARS

Although the honor or disgrace of being SF's primary ancestor is a debate that can never be settled finally, there can be no question that the rocket ship is the genre's primary icon. Even now when *Apollos*, *Titans*, and *Challengers* have appeared on the evening news blasting off from Cape Canaveral, the image of the rocket—preferably a '50s model kind with Pontiac tail fins—remains the sci-fi image of preference. It is there at the beginning of every *Star Trek* episode. It waltzes to the strains of the "Blue Danube" in *2001*. It is an identifier, like the cross or the hammer and sickle, with a single all-encompassing meaning, one that transcends all distinctions of class, taste, or even logic. *The Oxford Book of Science Fiction Stories*, published in 1992 by the most august of scholarly presses, features a simple hubcap-style flying saucer spinning through a cheerful blue void striped with a retro spectrum of oranges, greens, and yellows. Routledge, another scholarly press, presents *Reading by Starlight: Postmodern Science Fiction* (1995), a modishly structuralist critique by Damien Broderick, with a spaceship foregrounded against an india ink sky decorated with as many moons and planets as a Christmas tree. SF novels and collections with no thematic connection to space travel are nevertheless ornamented with spaceships and planets (Saturn by preference). My own novel, *334*, was paperbacked in England with a perfunctory spaceship/starry sky, despite the fact that in the future imagined in that novel, the space program has long since been abandoned for budgetary reasons and is only a source of nostalgia. The same spaceship, rotated some 120 degrees, was featured on the cover of the English edition of my anthology, *The New Improved Sun: An Anthology of Utopian SF* (1976), although only one of the fifteen stories ventures from the planet Earth. Quite simply, for publishers, as for the public at large, SF = spaceships + planets.

The reason may be simply that spaceships and planets are easy to illustrate. With a compass, a ruler, and limited experience with an airbrush or computer graphics software, any novice can produce a dozen such covers in a lazy afternoon. And such covers do the job. SF readers do not expect them to represent the contents of the books they read, only to signify the genre. In themselves, spaceships excite few readers. Little attention is given in most contemporary SF to the mechanical specifications of spaceships or the character of astronomical bodies viewed at telescopic distance. *Star Trek* episodes or tie-ins are rarely about the possible technologies of faster-than-light travel. Indeed, such technologies would seem not to be possible in an Einsteinian universe, and so all of the SF rationales are no more than fast talking. Got a field imbalance problem? It could be the Heisenberg compensators, so check out the pattern-buffer diagnostics.

This kind of con artistry has been going on since Edgar Allan Poe sent his Hans Pfaall to the moon in a hot-air balloon, which is constructed from "five iron-bound casks, to contain about fifty gallons each, and one of a larger size; six tin tubes, three inches in diameter, properly shaped, and ten feet in length; a quantity of a *particular metallic substance*, or *semi-metal*, which I shall not name, and a dozen demijohns of *a very special acid*. The gas to be formed from these later materials is a gas never yet generated by any other person than myself." If Poe had known about Heisenberg compensators, he would certainly have used them, too.

The first science-fictional voyage into outer space that attempts to put a high-tech gloss on the endeavor is found in Jules Verne's romance of 1865, *From the Earth to the Moon.*[1] Verne has been sniffed at for having his space travelers propelled by a gigantic cannon, since the acceleration required for take off would have squashed them flat. But in other details of his story, he has a prophetic edge on even such latecomers as

Robert Heinlein, particularly in the matter of the social engineering of the project. Heinlein, in the spirit of Ayn Rand, posited a single heroic capitalist, Delos D. Harriman, as responsible for the first flight into space and the colonization of the moon—despite the resistance of a "damn paternalistic government." Verne, though he does not assign the task to the official U.S. government, sees it as the accomplishment of a nascent U.S. military-industrial complex, which had "signally distanced the Europeans . . . in the science of gunnery. Not, indeed, that their weapons retained a higher degree of perfection than theirs, but that they exhibited unheard-of dimensions, and consequently, attained hitherto unheard-of ranges." A group of artillery professionals, disillusioned by the prospect of peace following the Civil War, organizes itself as the Gun Club, an organization of "1,833 effective members and 30,565 corresponding members," which sounds like a hybrid of the National Rifle Association and NASA. The Gun Club transforms their gigantic cannons not into plowshares but into spaceships, and the story begins.[2]

This is uncannily close to how NASA got going in the wake of World War II, using the talents of the German rocket scientists who developed the V-2 rockets. The space program has always been, in a functional sense, a work-relief program for the military during periods of peace. As Verne sums it up:

> One day, however—sad and melancholy day!—peace was signed between the survivors of the war; the thunder of the guns gradually ceased, the mortars were silent, the howitzers were muzzled for an indefinite period, the cannon, with muzzles depressed, were returned into the arsenal, the shot were repiled, all bloody reminiscences were effaced; the cotton plants grew luxuriantly in the well-manured fields, all mourning garments were laid aside, together with grief; and the Gun Club was relegated to profound inactivity.

Verne was, of course, French, and the French have their own distinctive tradition of science fiction. But in both his chief formative influence (Poe) and in his own forming influence on the genre, he may be reckoned an American *manqué*. Twenty-three of his sixty-four novels are set in an America that increasingly comes to resemble the nightmarish, capitalist dystopias of Bertolt Brecht's *Mahagonny* or Jack London's *The Iron Heel.*

In *Robur the Conqueror,* written in 1866, a year after *From the Earth to the Moon,* Verne portrays a heroic American capitalist in the Heinlein mold, who invents the airplane. In a sequel of 1904, *Master of the World,* the Byronic Robur has become a mad scientist and proto-Hitler. Air travel (not to mention space flight) has always been dangerous.

This darker side of Verne has generally been ignored or slighted in order to maintain a convenient dichotomy between himself and his usual antithesis, H. G. Wells. Wells is reputed the pessimist, a creator of nightmarish futures; Verne, the technophile optimist. But this does a disservice to both writers. Wells was also a notable utopist, and Verne, from the very beginning, had reasonable dreads for the new technological day adawning. However, as we now know, with the discovery of his suppressed manuscript, *Paris au XXe Siecle* (written in 1863, published only in 1994), Verne was not the unmitigated technophile of his Pantheon portrait. In *Paris* he does predict gas-powered automobiles (and traffic jams), fax machines and telephones, even the electric chair as an advance on the guillotine. But he also foresaw that the French language would be overwhelmed by English, and the poet-hero of his novel would search bookstores, hopelessly, for the works of Victor Hugo—exactly the dire fate that French intellectuals still profess to dread. Fortunately, for Verne's commercial success, his publisher and guardian angel, Jules Hetzel, nixed *Paris* and steered his pliant author on the course that made him famous, writing a succession of popular adventure novels, the *Voyages extraordinaires,* which made him immortal: *20,000 Leagues under the Sea, Around the World in Eighty Days, Journey to the Center of the Earth,* and the others, works whose work made Verne a name that Hollywood can conjure with, but that no one reads, at least in English.

The fact remains that in the year 1865, Verne wrote the first intentionally science-fictional account of a trip to the moon. Why the moon? There are some obvious answers. It is the nearest destination in outer space and, unlike the planets, it looks like a Somewhere. As long ago as the second century A.D., when Lucian of Samasota wrote his *True History,* it figured as an imaginary destination for adventurous travelers. To understand the phases of the moon was to realize that it was a sphere, like the earth, reflecting the sun's light. The distance could be calculated, and its size. It is immense, and it is always there, beckoning.

The planets are much more hypothetical. We can see them, and astronomers (after Copernicus and Newton) could infer their mass and motion, but not until the latter part of the nineteenth century, when telescopes allowed us to see (or, nearly as good, to imagine) canals on Mars, did the rest of the solar system begin to possess an interest equivalent to the moon's.

But the moon is not only *there,* as a physical presence; it is also a part of Heaven, and Heaven is much more than the sky that we can see at night. It is the abode that almost every culture has assigned the gods as their residence and the souls of the blessed as their dormitory. Therefore, to go to the moon is, literally, to set foot on Heaven's soil. The moon is featured in exactly that way in Dante's *Paradiso;* it is the first step up from Earth's Mount Purgatory, the antechamber to the farther planets and the Empyrean beyond. And *if* the moon is the beginning of Heaven, then isn't it fair to assume that there must be someone there? Inhabitants. People like us, but different. Aliens.

Enter the greatest science-fiction writer of them all (though the label had yet to be invented when he wrote the works we remember), H. G. Wells. Wells may have read Poe, but there is no evidence of any influence in what he wrote. He surely had read Verne but rarely emulated him. The marvelous quality of the SF novels and short stories he wrote between 1895 *(The Time Machine)* and 1914 (*The World Set Free,* in which he predicted the atomic bomb) is their masterful blurring of the boundaries between the two hyphenates of the genre, science and fiction. Verne's "science," when it is not simply antiquated and wrong, is that of the classroom. Reading a Verne novel can be like a visit to a provincial natural history museum, with cases and cases of specimens. In ten minutes, you want to head for the cafeteria.

Wells, by contrast, was a born novelist. In that, of course, he stood on the shoulders of giants. By the time he came to write his first books, in his thirties, he had absorbed the protocols and strategies of a century and a half of English novel writing. As a result, his best books can still be read with simple vicarious pleasure, and without that exercise of sheer willpower that Verne and (all too often) Poe demand.

Wells was also a formidable polemicist.[3] In the course of time, that would be his undoing. He became, even more than Verne, a Schoolteacher Absolute, a fate that would befall so many later SF writers—Heinlein, Asimov, Bradbury, Le Guin, Delany—that it must be considered an occupational hazard. But in his heyday, Wells combined both knacks—the fictive and the polemical—to produce works of modern mythology, their agenda so potently *imagined* that they overcame any skeptical reservations by their sheer narrative power.

The agenda that Wells advanced in these novels—*The Time Machine, The Island of Doctor Moreau, The War of the Worlds,* and *First Men on the Moon*—was Darwin's theory of evolution as applied to human history. Wells, before he came to write fiction, was a student of T. H. Huxley, Darwin's most effective champion. Huxley's formal debate in 1860 with Bishop Samuel Wilberforce was considered a knockdown victory for Darwinism, even by the Wilberforce forces. But that didn't much matter. People will believe what they want to believe. Evolution remains controversial. Indeed, Darwin's theory of evolution remains the essential fracture line in "modern" culture. And it was Wells, much more than his mentor Huxley, who tipped the balance in favor of the acceptance of the evolutionary hypothesis by creating myths sturdy enough to persist into our own time. A theory can be controverted; a myth persuades at gut level.[4]

But what has all this to do with rockets to the moon? Evolution evokes images of cavemen and dinosaurs—creatures of millennia past. Rocket ships and astronauts are emblematic of the future. And yet there is a common link between them, and that link can most clearly be discerned in the work of H. G. Wells, who realized that they were complementary images, representing equidistant points from the present on the evolutionary continuum. The anxieties provoked by Darwinian nature, in which mankind is only a superior kind of ape, are dispelled by the vistas of outer space, through which the whilom ape can soar in his new aspect as a "superman." The word was coined by George Bernard Shaw, in 1903, as a translation of Nietzsche's *übermensch,* who is conceived to be a more fully evolved form of homo sapiens. In his later, devolved condition as a comic book hero, Superman is nothing less than a rocket ship in human form, a man of steel able to fly by sheer willpower anywhere in the universe and equipped with X-ray vision and all kinds of internal *Star Wars* weaponry.

Almost all of Wells's best SF has an evolutionary subtext. In his first novel, *The Time Machine* (1895), the human race of the year 802,701 has bifurcated into distinct species: an upper crust of twitty Elois and the subterranean-dwelling, predatory Morlocks. An "Invention," Wells subtitled the book, and most readers have accepted his nightmarish vision of the existing Britsh class system as good escapist fun, but the same invention, when treated as an actual possibility, has been the source of one of the bitterest "battles of the books" of recent times. Richard Herrnstein and Charles Murray's *The Bell Curve*[5] hypothesized the same rigidification of class differences into genetically distinct under- and overclasses. The result was a cry of havoc and heresy. Even without its potential for racial divisiveness, *The Bell Curve*'s thesis dismays for the same reason that *The Time Machine* or *The Origin of Species* do: it suggests that human nature is not a God-given absolute, but an un-Platonic variable, the function of our malleable genes. We have been apes, and might become apes again.[6]

In his next SF novel, *The Island of Doctor Moreau* (1886), Wells puts a different and more horrific spin on evolutionary theory. Doctor Moreau is a latter-day Victor Frankenstein, who uses vivisection to create "beast men," human hybrids of leopards, pigs, dogs, monkeys, and other animals. His beast men exemplify both what is bestial in human nature and what is pitiful and even "spiritual" in animal nature—and the reason that Victorians of religious sensibilities recoiled from Darwinism with its vistas of a nature red in tooth and claw inhabited by a human animal stained by that blood.

Darwinian man is a beast not only in his origins but in his destiny. Then as now, the English liked their beef rare and their game high. The British Empire, like all others before it, represented the successful predation of the weak by the strong, and that was the subject, in a metamorphosed form, of *The War of the Worlds* (1898). Less ostensibly a Darwinian parable than its predecessors, it nevertheless carries on the same thread of thought. The invading force of Martian spaceships devastates London and the English countryside as effortlessly and irresistibly as a company of modern artillerymen might subjugate Samoa, or as the conquistadores subjugated Mexico and Peru. The parallels are not unintended, and they are given a distinctly Darwinian overlay in *The War of the Worlds*.

Just as Poe's Vankirk vatically proclaims in "Mesmeric Revelation" that "you will have a distinct idea of the ultimate body by conceiving of it to be entire brain," so Wells describes his Martians as "heads—merely heads. Entrails they had none." The reasons for this, he explains, are evolutionary. The Martians, living on an older planet, have had a longer time to evolve, and evolution has whittled them down to a predator's sine qua non, a brain. Wells reasons

> that the perfection of mechanical appliances must ultimately supersede limbs; the perfection of mechanical devices, digestion; that such organs as hair, external nose, teeth, ears, and chin were no longer essential parts of the human being, and that the tendency of natural selection would lie in the direction of their steady diminution through the coming ages. The brain alone remained a cardinal necessity.

Wells's Martians, thus, are the first bionic men, the primal Robocops, pure intelligence sheathed in imperishable steel, as fearful as they are, by evolutionary logic, inevitable.

Wells's most direct, nonmetaphorical diorama of the world according to Darwin is found in a mordant novella written in 1897, just after *Doctor Moreau*, "A Story of the Stone Age." Here, eighty-three years before Jean Auel's compelling saga kickoff, *The Clan of the Cave Bear*, Wells tells a tale of daily life in Stone Age England that is a miniature version of Auel's epic told with a drollery that even the likes of Henry James might appreciate. It begins where Darwin begins, with an appreciation of the broader perspectives of Earth's history that the science of geology made possible:

> This story is of a time beyond the memory of man, before the beginning of history, a time when one might have walked dryshod from France (as we call it now) to England, and when a broad and sluggish Thames flowed through its marshes to meet its father Rhine, flowing through a wide and level country that is under water in these later days and which we know by the name of the North Sea.

Wells continues in this vein awhile, evoking an earlier, alternate England populated by the familiar—swallows, white ranunculus, lady's smock—

but also by hippopotami, lions, elephants, and "a number of little buff-colored animals" who are our ancestors. There is also, as in Auel's novel, a cave bear—two of them—but Wells, taking a holiday from "serious" fiction, grants his cave bears the gift of speech, as Kipling did in his *Jungle Book*. Their domestic dialogue, when they've returned to their cave after a humiliating encounter with the First Axe, wielded by the story's hero, Ugh-lomi, is one of Wells's drollest and most resonant inventions. Andoo, the husband, speaks first:

> "I never was so startled in my life. . . . They are the most extraordinary beasts. Attacking *me!*"
>
> "I don't like them," said the she-bear, out of the darkness behind.
>
> "A feebler sort of beast I *never* saw. I can't think what the world is coming to. Scraggly, weedy legs. . . . Wonder how they keep warm in winter?"
>
> "Very likely they don't," said the she-bear.
>
> "I suppose it's a sort of monkey gone wrong."
>
> "It's a change," said the she-bear.
>
> A pause.

The bears speculate about the new beasts' use of weapons and fire—"that glare that comes in the sky in daytime—only it jumps about." He's lazily curious. She has other matters on her ursine mind:

> "I wonder if they *are* good eating?" said the she-bear.
>
> "They look it," said Andoo, with appetite—for the cave bear, like the polar bear, was an incurable carnivore—no roots or honey for *him*.
>
> The two bears fell into a meditation for a space. Then Andoo resumed his simple attentions to his eye [which was wounded by Ugh-lomi's axe]. The sunlight up the green slope before the cave mouth grew warmer in tone and warmer, until it was a ruddy amber.
>
> "Curious sort of thing—day," said the cave bear. "Lot too much of it, I think. Quite unsuitable for hunting. Dazzles me always. I can't smell nearly as well by day."
>
> The she-bear did not answer, but there came a measured crunching sound out of the darkness. She had turned up a bone. Andoo yawned.

> "Well," he said. He strolled to the cave mouth and stood with his head projecting, surveying the amphitheater.

The reader knows that poised at the top of the cliff above Andoo's massive, sleepy head is the boulder that the cave man has positioned to smash the cave bear's skull. Indeed, all through Andoo's ruminations, we know that boulder is there, so that the bear's every word resonates with tragicomic irony. He is like the outsized oaf in a Chaplin one-reeler whose pratfalls we gleefully anticipate. But he is also a representative Victorian paterfamilias trying to come to terms with these new, outlandish ideas by that fellow Darwin. He is the old world about to be crushed by the new. And when the inevitable does happen and Andoo's corpse lies decomposing and gnawed by hyenas at the mouth of the cave, Wells shifts the tone from farce to elegy:

> The she-bear stopped dead. Even now, that the great and wonderful Andoo was killed was beyond her believing. Then she heard far overhead a sound, a queer sound, a little like the shout of hyenas but fuller and lower in pitch. She looked up, her little dawn-blinded eyes seeing little, her nostrils quivering. And there, on the cliff edge, far above her against the bright pink of dawn, were two little shaggy round dark things, the heads of Eudena and Ugh-lomi, as they shouted derision at her. But though she could not see them very distinctly she could hear, and dimly she began to apprehend. A novel feeling as of imminent strange evils came into her heart.

It the mark of a great writer, as against a hack, that he invests his villains with a humanity equal to that of his heros, and that is what Wells does in this story. Ugi-lomi and his mate, Eudena, are vividly human, a proper Adam and Eve, but like that other pair, they are also nasty little upstarts. Andoo and his mate, for all their conversational complacencies, are the poetic fulcrum of the story. They are evolution's losers.

And they are also the Victorian middle class in a nutshell, with "dawn-blinded eyes" that cannot foresee their own needful extinction—needful, that is, if humanity is to begin its triumphal march across the planet.

Wells also sent his own expedition to the moon. He had to, really, for

Verne had set a precedent in *De la terre de la lune* and its routine 1870 sequel, *Autour de la lune*,[7] and *The First Men in the Moon* represents Wells's submission to the demands of genre authorship. Sir Edmund Hillary climbed Mount Everest because it was there, and Wells went to the moon in much the same spirit.

Even so, it is a spirited flight of fancy propelled by the auctorial fiat of "cavorite." Gravity is the enemy of space flight, so Wells invented antigravity—for which Verne, still alive when Wells's book came out in 1901, was implacably disdainful: cavorite was not *scientific*. Like the cave bear Andoo, Verne had had a whiff of the future, and it did not appeal to him. Wells was a nimbler novelist, able to mix keen social satire with adventure and give the whole thing a gloss of psychological depth. His works were admired by the literati of his day, while Verne's had already been relegated to "boys' adventure." The same kind of rivalry has continued to the present.

But Verne and Wells were united in one thing: that the moon seemed a thrilling and *possible* destination. Their contemporaries might have regarded their accounts of voyaging to the moon as mere fancies, in the tradition of Lucian or Cyrano. Both authors knew better. Verne's novel has passages that a NASA publicist might quote with little emendation, while Wells's hero, Bedford, though portrayed as a kind of lunar '49er lusting for gold, discovers a vein of idealism in his own soul when he actually arrives on the moon and wonders why he'd come there. "What is this spirit in man that urges him for ever to depart from happiness, and security," he asks himself, "to place himself in danger, even to risk a reasonable certainty of death? It dawned upon me there in the moon as a thing I ought always to have known, that man is not made simply to go about being safe and comfortable and well fed and amused. Against his interest, against his happiness he is constantly being driven."

When Wells' voyagers alight from their spacecraft, they discover something that Verne, the rationalist, was too proud to purvey for his readers: aliens. Wells's Selenites are a more developed variation of his Martians—not *all* brain, but tending in that evolutionary direction. At the end of a ripping tale, after being held captive by the superrational Selenites and escaping with a fortune in gold (the moon's commonest metal, happily for Bedford), Wells drives home his essential point:

mankind stands at an evolutionary turning point between the carnivorous ape and the soulless Selenite, between anarchic brutality and and an orderly but totalitarian scientific civilization.

It is where humanity has stood ever since—and must, for the foreseeable future. The same antithesis will take center stage again and again: in Aldous Huxley's *Brave New World* (whose denizens are human Selenites, genetically modified to fit their predestined roles in the labor force) and no less memorably in Clarke's and Kubrick's *2001: A Space Odyssey* (1968), in which a bloodstained First Axe, very much like that of Ugh-lomi's, metamorphoses into a spaceship that is guided by a being who is pure mentation, HAL. There is a further, a complementary metamorphosis, as the human spaceman David Bowman is transformed into something altogether superhuman. Here is Clarke's account of the transforming moment in the novel version of the tale (which explains much that seems obscure in the movie's visually stunning denouement):

> The metal and plastic of the forgotten space pod, and the clothing once worn by an entity who had called himself David Bowman, flashed into flame. The last links with Earth were gone, resolved back into their component atoms.
>
> But the child [the metamorphosed Bowman] scarcely noticed, as he adjusted himself to the comfortable glow of his new environment [the vacuum of space]. He still needed, for a little while, this shell of matter as the focus of his powers. His indestructible body was his mind's present image of itself; and for all his powers he knew that he was still a baby. So he would remain until he had decided on a new form, or had passed beyond the necessities of matter.[8]

This defining moment in the history of SF is not without precedent. Besides Madame Blavatsky's insistence on such a future evolutionary leap into the sublime, there is a more obvious precedent, found in Clarke's own novel of 1953, and his first enduring contribution to the genre, *Childhood's End*, at the conclusion of which the children of Earth "evolve," in one planet-annihilating blast of "inconceivable metamorphosis," and unite with the cosmic "Overmind." The Luciferian "Overlords" who superintend humanity's metamorphosis are like Wells's

Selenites in being all intellect, and so lacking the capability for evolving into an immaterial union with the Overmind. Undoubtedly, when Kubrick approached Clarke to script *2001*, he already had envisioned a replay of that climactic set piece, so that the movie (and novel) can fairly be considered a recension of the earlier work.

But there are earlier precedents as well, of which the most grandiose is *Star Maker* by Olaf Stapleton (1937). Stapleton, like Clarke, is an English epigone of H. G. Wells. Stapleton's Star Maker is a disembodied Englishman of the '30s who takes off on a voyage of the universe, witnesses all kinds of aliens, and finally consorts with various immaterial essences, including the Creator. It is a Vision more than a novel, and its impact did not extend far outside the walls of the SF ghetto. Yet it remains the purest specimen of a rarefied category—the SF High Sublime—a book of undeniable merit but with the chill, and the longueurs, of a night outdoors in freezing weather watching the aurora borealis.

Wells, Clarke, Huxley, Stapleton: rather a large English contingent for a genre that I would claim to be *essentially* American. And it must be said, to their credit, that until the early 1950s, the English had a virtual monopoly on the writing of enduring SF. In America, since Poe and until the '50s, there were only a few utopian and dystopian novels of note, and those are of interest chiefly to readers of a scholarly bent: Edward Bellamy's *Looking Backward* (1888), Ignatius Donnelly's *Caesar's Column* (1890), and Jack London's *The Iron Heel* (1907). Americans were too busy building the future to bother imagining it.

But Poe has staked his own claim on outer space as well, having scooped Clarke in the matter of global, planetary transcendence, and in a manner distinctively American, as we shall see. Poe actually offers two clear precedents of planetary/racial metamorphosis: his "Mesmeric Revelation," noted above, and "The Conversation of Eiros and Charmion," which begins in a tone of uncompromising *haut faux*, thusly: "Eiros: 'Why do you call me Eiros?' Charmion: 'So henceforward will you always be called. You must forget, too, *my* earthly name, and speak to me as Charmion.'" As it turns out, they have been *transformed* into Eiros and Charmion by the close passage of a comet, and *a total extraction of the nitrogen* (Poe's italics) from the Earth's atmosphere. The result, as Eiros recalls, was

> irresistible, all-devouring, omni-prevalent, immediate; the entire fulfillment in all their minute and terrible details, of the fiery and horror-inspiring denunciations of the prophecies of the Holy Book.
>
> . . . A furious delirium possessed all men; and, with arms rigidly outstretched toward the threatening heavens, they trembled and shrieked aloud. But the nucleus of the destroyer was now upon us; even here in Aidenn, I shudder while I speak. Let me be brief—brief as the ruin that overwhelmed. For a moment there was a wild lurid light alone, visiting and penetrating all things. Then . . . the whole incumbent mass of ether in which we existed burst at once into a species of intense flame, for whose surpassing brilliancy and all-fervid heat even the angels in high Heaven of pure knowledge have no name. Thus ended all.

Does that remind you of any particular contemporary event? Hiroshima, perhaps? We'll get *there* in the next chapter. Meanwhile, Poe himself points, with a loud shriek and rigidly outstretched arm, to the oldest and clearest precedent of all, Paul's First Epistle to the Corinthians, Chapter 15, verses 50 to 55:

> Now this I say, brethren, that flesh and blood cannot inherit the kingdom of God; neither doth corruption inherit incorruption.
>
> Behold, I show you a mystery; We shall not all sleep, but we shall all be changed,
>
> In a moment, in the twinkling of an eye, at the last trump: for the trumpet shall sound, and the dead shall be raised incorruptible, and we shall be changed.
>
> For this corruptible must put on incorruption, and this mortal must put on immortality.
>
> So when this corruptible shall have put on incorruption, and this mortal shall have put on immortality, then shall be brought to pass the saying that is written, Death is swallowed up in victory.
>
> O Death, where is thy sting? O grave, where is thy victory?

That passage is one among a few crucial biblical texts that are the basis, among fundamentalist millenarians, for the notion of the Rapture: that moment, according to believers, prior to the more dire events of the Apocalypse, when the faithful are lifted up, en masse and all at once, into the sky, to greet the Messiah and share with him a ringside view of the prophesied and inevitable end of "the Late Great Planet Earth."

This notion is something of a theological novelty. Since Augustine's time, when it had already begun to appear that the Second Coming might still be a long time coming, theologians have tended to look on the apocalyptic prophecies of Revelations with an allegorical eye. By the Victorian era, the elaborate scenario set forth in the Bible had been whittled down, for practical purposes, to an a-millennial view: a Second Coming of an undetermined date, followed by Eternity. The seven seals, the Anti-Christ, the Whore of Babylon, the Battle of Armageddon, and all those other exciting things were relegated to the status of visionary pronouncements or, simply, fiction.

But then came Darwin, and among the many responses to Darwin (besides SF) was a new interest in, and insistence on, the Bible's own *future* histories. To defend the Bible in its account of our origins—Adam and Eve, the Tower of Babel, the Flood—became, increasingly a rearguard action. The alternative was transcendence, a flight (like science fiction's) into the future; corruption putting off incorruption, just as it does in Paul's Epistle, or in *Childhood's End.*

But what about the *moon?*

Hal Lindsey has the answer to that. Hal Lindsey is the author of a number of best-selling millenarian tracts, beginning with *The Late Great Planet Earth* (1970) and continuing ever since, due to the continued existence of both this planet and Hal Lindsey. In *The Rapture* (1983) Lindsey, after quoting verse 52 above ("In a moment, in the twinkling of an eye . . ."), makes the following commentary:

> Someone said that the twinkling of an eye is about one-thousandeth of a second. The Greek word is *atomos* from which we get the word atom. It means something that cannot be divided. In other words, the Rapture will occur so quickly and suddenly that the time frame in which it occurs cannot be humanly divided.

Just think of it . . . in the flash of a second every living believer on earth will be gone. Suddenly, without warning, only unbelievers will be populating planet earth.

I recently watched in awe as the space shuttle blasted off into space. Within a matter of a few minutes it was out of sight and traveling at more than six times the speed of sound. What will take place for each living believer at the Rapture surpasses this by all comparison.[9]

There, with a precision that only the unconscious mind can command, is the connection: Blast-off = the Rapture.[10] The Rapture and its associated pleasures have sold not only many millions of books for Lindsey, but it has inspired such memorable end-time kitsch as "Rapture wrist watches with the words 'ONE HOUR NEARER THE LORD'S RETURN' inscribed around the face; bumper stickers proclaiming 'Beam Me Up, Lord,' or 'Warning: If the Rapture Occurs, This Car Will Be Driverless'; and full-color Rapture paintings, complete with crashing cars and planes, available in postcard format, as framed prints, and as laminated placemats."[11]

As those bumper stickers suggest, the Rapture is, in some respects, an in-your-face, middle-finger challenge to the world at large. Even if one does not expect to be translated momentarily into the ether, what fun to share the double entendre with the driver right behind you. It may be that among most Americans today, beliefs are "entertained" in much the same spirit that other cultural products are enjoyed. That would explain the perennial popularity of millennialist belief despite the repeated failure of Doomsday to materalize at its appointed hour. As a fantasy, the Rapture can be replayed over and over, like a favorite song. One doesn't require a song to be "true." It just has to have a tune you like to hum.

Here is the basic melody as delivered by Ray Bradbury, in a 1979 interview:

Orwell's *1984* came out 30 years ago this summer. Not a mention of space flight in it, as an alternative to Big Brother, a way to get away from him. That proves how myopic the intellectuals of the 1930s and 1940s were about the future. They didn't want to see something as exciting and soul-opening and as revelatory as space travel. Because we

> *can* escape, we *can* escape, and escape is very important, very tonic, for the human spirit. We escaped Europe 400 years ago and it was all to the good.[12]

If Bradbury's emphasis in the leitmotif "we can escape" fell on the first word instead of the second, its meaning would be clearer. For the true-believing SF fan, as for religious millenarians, the promise of liftoff from Earthside is reserved for a privileged and superior few. The simpler sort of SF fan, the kind who attends *Star Trek* conventions dressed in Enterprise-style pajamas, believes the moon is his destination. Heinlein, in his early fiction, promoted the idea of the moon as an off-planet clubroom for aspirants to Mensa. In "It's Great to Be Back," a story that appeared in the *Saturday Evening Post* in 1947, the hero explains why he's decided to return to his job on the moon: "I found out I was a Loony at heart quite a while ago. . . . My feet hurt [because of Earth's higher gravity], and I'm sick of being treated like a freak. I've tried to be tolerant but I can't stand groundhogs. I miss the civilized folks in dear old Luna."

In Bradbury's rhetoric and in Heinlein's early stories, outer space figures as the Next Frontier, somewhere to head once the westward course of empire has reached its limit at the Pacific coast, somewhere beyond the prying eye of Big Brother and the other constraints of an over-civilized, over-regimented, over-taxed existence.

This is an illusion. The average citizen, now and in the foreseeable future, has a much better chance of being elected to the College of Cardinals than of flying to the moon or beyond. As Arthur Clarke spells it out in *Profiles of the Future*, a nonfiction examination of SF's most popular misconceptions:

> The Space Frontier is infinite, beyond all possibility of exhaustion; but the opportunity and the challenge it presents are both totally different from any that we have met on our own world in the past. All the moons and planets of this Solar System are strange, hostile places that may never harbour more than a few thousand human inhabitants, who will be at least as carefully handpicked as the population of Los Alamos. The age of mass colonization has gone forever.[13]

What Clarke is too polite or too politic to point out is that the hand that has done the picking for the space program is Big Brother's. Given the complexity, fragility, and vulnerability of the whole enterprise, matters could not be ordered otherwise.

In the immediate postwar era, when the space program was initiated, Big Brother did not need to create a climate of opinion favorable to funding the space program. The Cold War was soon in high gear, and the U.S. military establishment had carte blanche for scientific research. The lion's share of that budget was devoted to atomic weaponry, but full-scale thermonuclear war would require intercontinental missiles, and a space program was funded accordingly. Some of the scientists recruited for this effort, such as Wernher von Braun, who developed the V-2 rocket for the Nazis, had a sulfurous smell about them, but until 1957, when the Russians launched *Sputnik*, the space program didn't have to worry about its image. It was supposed to have been a top secret.

After *Sputnik* and further Soviet successes, the space program began to work on its persona. Tom Wolfe's *The Right Stuff* describes the post-*Sputnik* recruitment and grooming of the first astronauts. Wolfe huffed and he puffed, but he could not inflate the astronauts into anything more than a superior sort of jock. Their dialogue, on the page and in the movie version of 1983, is as leaden as an NFL locker-room interview: "What I think is that the whole team has got to pull together, and then we'll go out there and win!" Hollywood's next big favor to NASA was *Apollo 13* in 1995, a movie that shows that if the whole team pulls together, they're going to win.

This is a worthy moral to inculcate, and many of the astronauts might well be interesting people in real life. But as astronauts, they were dull. With the best soundtrack and cinematography money can buy, Hollywood has not been able to get around the fact that the astronauts are as little in charge of their own destiny as people on a roller coaster. The movie *Apollo 13* offers a situation in which the astronauts can offer some minuscule input toward their survival, but mostly they just have to hang on tight. That is the stuff not of heroism but of nightmares.

Not surprisingly, NASA is no longer beating the drum for manned missions to the moon. The moon was of use to NASA only while it was virginal dreamstuff. Once we had put some footprints on it, planted the flag in its dust, and taken some souvenir snapshots, additional landings did not excite much attention. As the TV ratings plummeted, Congress grew stingy. The deaths of seven astronauts in the explosion of the *Challenger* shuttle in 1986, and the subsequent finger pointings, left NASA looking like just another funding pipeline of the military-industrial complex, a minor annex of the Pentagon. Conceivably, the lunar dust has received its last human footprint.

If so, it will have little impact on science fiction, which had abandoned the moon at about the time *Apollo 11* landed there. As David Hartwell has remarked in his history of the genre, *Age of Wonders*, "When science fiction comes true, it's no fun anymore." *Sputnik* was an answered prayer, a certain source of disillusion:

> Some SF people have continued to believe in the romance of space travel throughout the 1960s and 1970s, in the face of the boring facts, but a lot of others, particularly sensitive to what had been their own province for decades, became more and more alienated from U.S. media/government space travel. What good is it to have predicted all this technology, from space suits and orbital velocities to stage rockets and communication satellites, if roles can be performed by trained chimpanzees? . . .
>
> . . . By the time of the second moon landing, all according to TV script and utterly anticlimactic, even the most committed SF people had begun to mutter and grumble that the right thing was being done by the wrong people in the wrong way. How could they make it so unromantic? How could they?[14]

The public indifference to, and even resentment toward, NASA and the prosaic realities of the exploration of space has been an occasional theme of SF. The 1977 film *Capricorn One* posits a faked flight to Mars that takes place on a film set. Twenty years later, the landing of *Pathfinder* on Mars and the adventures of little *Sojourner* have provoked rumors that all the CNN footage is also a hoax, which shows that even the lunatic fringe can be skeptical when that suits its agenda. If the real world denigrates their UFOs as mere imposture, then they will do the same for the planet Mars.

The English SF writer J. G. Ballard has been targeting NASA as the object of his deadpan satire for decades. His favorite image is that of Cape Canaveral's launching towers and NASA's other immense machineries lying in ruins, drowned cathedrals of the Age of Space. In "The Message from Mars," published in 1992 and one of his best stories in years, Ballard describes a manned mission to Mars that, though not an outright hoax, certainly qualifies as a fake, having been cast and scripted to conform to the demands of the TV audience. When the five astronauts return from their mission, a continued run is planned for the show:

> An unending programme of media appearances awaited the astronauts—there would be triumphal parades through a dozen major cities, followed by a worldwide tour lasting a full six months. NASA had already appointed firms of literary agents and public relations experts to look after the commercial interests of the astronauts. There were sports sponsorships, book contracts, and highly paid consultancies. . . . Two days before the Zeus IV landed, NASA announced that three major Hollywood studios would collaborate on the most expensive film of all time, in which the astronauts would play themselves in a faithful recreation of the Martian voyage.[15]

However, when the *Zeus IV* does land, the astronauts refuse to come out—and to explain their refusal. Nothing short of a nuclear explosion can penetrate the thermal plating of the hull, and the ship is provisioned for forty years. While the public speculates, the uncooperative astronauts carry on their domestic routines with Zen-like imperturbability—until they die, decades later.

The public's disenchantment with Mars, like its earlier loss of interest in the moon, takes diverse forms, many of which found expression in the Sunday supplement of my local newspaper, the Middletown, New York, *Times-Herald Record,* devoted to the Pathfinder mission. A feminist columnist reacted with a big shrug: Mars is a guy thing. Half a page was devoted to the Hot Action Pack JPL Sojourner Mars Rover, available for only $5 from Mattel, which will allow people to make their own hoax footage of Mars landings at a local gravel pit. Then there was a story on the local barber, Butch Hunt, who has seen UFOs himself and gets written up in the *Record* any time outer space hits the news. Video rentals of SF movies about Mars were graded, with *The War of the Worlds* getting higher points than the recent *Mars Attacks*. The overall impression was that Mars was strictly dog-days trivia. A later news story on a meeting of local amateur astronomers who'd gathered to discuss the *Pathfinder* mission ended in this wise: "None of the 20 or so people in the audience seemed to find the discussion the least bit boring. 'As Carl Sagan used to say, 'It's a wonderful time to be alive," Lindemann said."

Carl Sagan, who died just months before the *Pathfinder* mission and the release of *Contact*, the movie based on his SF novel about contact with extraterrestrials, is a sadly emblematic figure. He was consistently one of NASA's biggest promoters and actively involved in many of its projects. His angle was always that there *must be* life out there. In pursuit of that faith, he helped to organize SETI, the Search for Extraterrestrial Intelligence. SETI began to broadcast messages to the stars, hoping that someone would beam a message back. To date there has been no reply. Sagan's response to the silence of the stars was to invent one. That, in a nutshell, is the plot of *Contact*. Sagan, who throughout his career had been a principled debunker of UFO fanatics, will probably be remembered best for having assisted in buttressing that faith.

In the time of Lucian of Samasota, or even of Cyrano de Bergerac, a trip to the moon was the stuff of fantasy. Once such a voyage began to seem a concrete possibility, science fiction established itself as a separate genre, but the tropism toward fantasy remained in its genes, and now that the moon and Mars and most of our solar system have come to seem real—but barren—destinations, SF has reverted to its origins as fantasy. For every SF story that posits interstellar travel and adventures among aliens is a trip to Oz, given what we know of interstellar distances and the constraints of relativity theory.

As Robert Heinlein observed, in another context, the moon is a harsh mistress.

Chapter 4

HOW SF DEFUSED THE BOMB

On the day Hiroshima was bombed, August 6, 1945, I was five years old. I have no memory of the event, or indeed of anything else connected with World War II. In the grade schools I attended from 1946 through 1953, there must have been civil defense exercises of the sort depicted in the archival footage of the 1982 documentary, *The Atomic Cafe*, in which Burt the Turtle advises schoolchildren about to be bombed to "Duck and cover!" (assume a fetal position under their desks), but I don't remember that either.

I *can* remember reading, at age twelve, Ray Bradbury's *The Martian Chronicles*, a collection that includes his 1950 nuclear parable, "There Will Come Soft Rains,"[1] which describes the domestic routines of an automated house in Allendale, California. No one is home. A breakfast is prepared and scraped into the sink. Robot mice suck up the dust. "The house," Bradbury explains, "stood alone in a city of rubble and ashes. This was the one house left standing. At night the ruined city gave off a radioactive glow which could be seen for miles." The house's former inhabitants can be seen, in silhouette on an outside wall, a nuclear family of Dad, Mom, son, and daughter, "their images burned on wood in one titanic instant." That image is erased when the house is destroyed by an accidental fire.

In the next, and final, story of the book, "The Million-Year Picnic," another nuclear family manages to survive by escaping to Mars on the last working rocket ship. As Dad explains the situation to his kids when they arrive on Mars, "Wars got bigger and bigger and finally killed Earth. That's what the silent radio means. That's what we ran away from."[2]

As a child of that time, I received such explanations of the theoretical basis of the atom bomb as were available at school and in the pages of popular science magazines. As a result, by age twelve, I knew as much about atomic energy as all but a small minority of U.S. citizens, since the government's monopoly of information on the subject was virtually absolute. Even the head of the Atomic Energy Commission, David Lilienthal, was wont to complain about the "growing tendency in some quarters [the military] to act as if atomic energy were none of the American public's business."[3] For the most part, however, Lilienthal toed the official government line, minimizing the dangers of atomic warfare and radiation and rhapsodizing over the utopia that cheap atomic power would bring about—such a utopia as is promised in *Operation Atomic Vision*, a handbook issued by the AEC in 1948 for use in high schools:

> You may live to drive a plastic car powered by an atomic engine and reside in a completely air-conditioned plastic house. Food will be cheap and abundant everywhere in the world. . . . No one will need to work long hours. There will be much leisure and a network of large recreational areas will cover the country, if not the world.[4]

Essentially, the national consensus was to see no evil, hear no evil, and speak no evil. "To ask questions in this field is unpatriotic," Herbert Marks complained in the out-of-the-way-pages of the *University of Chicago Law Review.*[5] Many would have endorsed that proposition without demur, for the government then commanded a respect from the general public that makes the late '40s look, fifty years later, like the Age of Faith. That same government, after all, had just won the most extensive war ever fought on the planet, and the climactic victory had been the result of scientific research that had been conducted secretly. The reasons for such secrecy still existed: no one would wish his enemy to have such a weapon.

But the seeds of fear had been planted. In the immediate aftermath of Hiroshima, newspapers printed maps of American cities overlaid by concentric circles of destruction, so that one might estimate one's own chances of survival according to the distance of one's home from Ground Zero. On August 25, 1945, the *New Yorker* published an interview with John W. Campbell, Jr., the editor of *Astounding Science-Fiction*, in which Campbell described the likely outcome of a full-scale nuclear war: "Every major city will be wiped out in thirty minutes. . . . New York will be a slag heap. . . . Radioactive energy . . . will leave the land uninhabitable for periods ranging from ten months to five hundred years, depending on the size of the bomb."

Just one year later the *New Yorker*, in its August 31, 1946, issue, was to publish the single most influential document of the atomic age, John Hersey's *Hiroshima*. In a level but not affectless tone, Hersey traced the individual destinies of six Hiroshima residents from the morning of August 6 through the hours and days following the blast. Through the eyes of Hersey's six protagonists, we begin to apprehend the extent of the catastrophe: the miles of ruins, the heaps of dead, and crowds of survivors dying, just hours after the blast, of radiation burns: "yellow at first, then red and swollen, with the skin sloughed off, and finally, in the evening, suppurated and smelly." The horrors are in some ways more nightmarish than those of Auschwitz, for they are the result not of a series of human actions, which are physically if not morally comprehensible, but rather of a single, instantaneous Let There Be Death. In the twinkling of an eye, 100,000 people were dead and a city in ruins.

The years of acquiescent silence that followed Hiroshima were not a time of ignorance or innocence. The bomb had been unveiled. Only by following the advice of Burt the Turtle—"Duck and cover!"—could one escape knowledge of its potency and its potential, and even then, hiding under the desk with one's eyes tightly closed, one was penetrated by a new kind of dread.

Surely it would not be wildly theoretical to liken this situation to a state of repression, which *Webster's Ninth* defines as "a process by which unacceptable desires or impulses are excluded from consciousness and left to operate in the unconscious." Repressed materials tend to resurface in the form of nightmares—or, at a cultural level, as pulp fiction and

B movies—in short, as sci fi, especially as the genre existed in that more innocent era, when even ostensibly adult entertainment often had a childlike quality. Once again, it is Ray Bradbury (a lifelong child impersonator of a stature equal to that of Pee-Wee Herman) who created the defining image in his 1951 story from the *Saturday Evening Post*, "The Fog Horn," which became the film *The Beast from 20,000 Fathoms* (1953): a slumbering dinosaur, awakened by a nuclear blast, mistakes the sound of a fog horn for a mating call and levels New York City (in the movie version) looking for love. Bradbury's account of how his tale came to be written is wonderfully suggestive:

> One night when my wife and I were walking along the beach in Venice, California, where we lived in a thirty-dollar-a-month newlyweds' apartment, we came upon the bones of the Venice Pier and the struts, tracks, and ties of the ancient roller-coaster collapsed on the sand and being eaten by the sea.
>
> "What's that dinosaur doing lying here on the beach?" I said.
>
> My wife, very wisely, had no answer.
>
> The answer came the next night when, summoned from sleep by a voice calling, I rose up, listened, and heard the lovely voice of the Santa Monica Bay fog horn blowing over and over and over again.
>
> Of course! I thought. The dinosaur heard that lighthouse fog horn blowing, thought it was another dinosaur arisen from the deep past, came swimming in for a loving confrontation, discovered it was only a fog horn, and died of a broken heart there on the shore.[6]

There, in a single gestalt that will be repeated in movie after movie for the rest of the century—from *Godzilla* in 1954 to *Jurassic Park* of 1993 and its sequel, *The Lost World*, of 1997—is the meaning of the Bomb stripped of all theory and science. The Bomb is simply the greatest of all Monsters from the Id. "They Couldn't Believe Their Eyes!" declared the poster for the movie. "They Couldn't Escape the Terror! And Neither Will YOU!" It is a force that levels cities, but it is also, like other monsters—Frankenstein's creature, Dracula, the Phantom of the Opera—looking for love.

Unfortunately for would-be peacemakers, bombs—and other

weapons of all sorts—have a built-in erotic component, and the atomic bomb, as the most powerful bomb of all time, has inspired almost everyone who has written or made movies about it to represent it as a kind of aphrodisiac releasing the forces of the Id. The very image of the expanding nuclear mushroom is a visual pun so blatant that alert censors should long ago have forbidden its use by the media, and it's an easy morph from there to Bradbury's long-necked amorous dinosaur.

But dinosaurs have other symbolic resonances. They are extinct—as we may soon be ourselves, thanks to the bomb. The sight that prompted Bradbury's tale was an image of Western civilization—an abandoned roller coaster—in ruins and collapsing into the ocean: America as the New Atlantis, mourned by that misapprehended fog horn, which was not, after all, a love call but a warning that went unheeded.

Other B-movie monsters released by nuclear tests or radioactivity include Godzilla's various challengers (Mothra, Gigan, Mechagodzilla, and the others), the giant ants of *Them!* (1954) and various other enlarged city-wrecking insects, and, inevitably, *The Amazing Colossal Man* (1957). Radiation creates mutants, and in due course the Bomb would be responsible for just about every mutation that a special effects team was capable of: men with terrible skin diseases or two heads, or simply bad-ass bikers of the Mad Max variety.[7]

I was in seventh grade when these first fun house–mirror reflections of nuclear dread began to appear, and I absorbed them into my young psyche with an avid and uncritical receptivity. I liked the taste and glutted myself—not only on sci-fi movies (for they were still few and far between, and the great recirculating fountain of television was not yet in perpetual motion) but also with the more developed and persuasive fictions of the SF magazines and early paperbacks.

Nuclear dread reached fever pitch in these years. Russia had had its own atom bomb since September 1949. Soon after that, research was begun on the H-bomb. The trials of Fuchs, the Rosenbergs, and Alger Hiss set the stage for McCarthyism. The Korean War began, and with it speculation on whether the Bomb should be used against North Korea and China.

Necessarily, the science fiction of that era was pervaded by the same dread. In the nature of nightmares, the situations and imagery shifted under the torsion of strong feelings. Not every story engendered by nuclear anxiety made specific reference to the bomb. Novels like George Stewart's *Earth Abides* (1949), Ward Moore's *Greener Than You Think* (1947), and John Christopher's *The Death of Grass* (1956), which chronicle the extinction of humanity by plagues or ecological catastrophe, are products, even so, of nuclear dread. Of the first thirty novels cited by David Pringle in his readers' guide, *Science Fiction: The 100 Best Novels* (1985), all published between 1949 and 1959, seven are directly concerned with nuclear war, its prevention or its aftermath,[8] while another eight are clearly legitimate progeny of the bomb.[9]

Among the most popular or critically acclaimed SF authors of the postwar years I can think of very few who have not wrestled once or twice with the angel of apocalyptic dread, but there are two who were preeminent in that arena. One was Robert Heinlein, whom a majority of fans would unhesitatingly name as the greatest SF writer of the century. Indeed, Heinlein was regularly voted "best all-time author" in polls of SF readers. His work has compelled the admiration not only of fandom but of critics who deplore his views and lament his influence. One such critic, H. Bruce Franklin, begins his book-length treatise, *Robert A. Heinlein: American as Science Fiction* (1980), by making a claim for the joint significance of Heinlein and of science fiction that I would consider megalomaniacal if I did not agree so much with it:

> The phenomenon of Robert A. Heinlein expresses, among other things, the extraordinary quality of the everyday experience of our century. Heinlein is certainly our most popular author of science fiction, easily the most controversial, and perhaps the most influential. And science fiction has moved inexorably toward the center of American culture, shaping our imagination (more than many of us would like to admit) through movies, novels, television, comic books, simulation games, language, economic plans and investment programs, scientific research and pseudo-scientific cults, spaceships real and imaginary.[10]

Heinlein's first parable of the atomic age scooped the SF competition (and reality) by a few years. Published in the May 1941 *Astounding*, when Heinlein was thirty-four years old, "Solution Unsatisfactory" lays out a scenario of future history in which the United States defeats Germany in World War II by achieving atomic superiority and then tries to establish "a military dictatorship imposed by force on the whole world." A Russian-led "Eurasian Union" resists this benign initiative, makes an atomic sneak attack on the United States, and gets nuked in turn. Finally, the hero of the story, Colonel Clyde Manning, graduates from his position as Commissioner of World Safety to become "the undisputed military dictator of the world," displacing a U.S. President who had proved unequal to the exercise of full imperial authority. In this single novelette written before Pearl Harbor, Heinlein anticipates not only the role of atomic power in bringing World War II to an end, but the ensuing nuclear standoff of the Cold War as well, and he concludes with a military coup d'état that prefigures the basic story line of two classic films of the High Cold War, *Dr. Strangelove* (1962) and *Seven Days in May* (1964), with the significant difference that the generals attempting the coups in those movies are pictured as lunatics rather than world saviors like Heinlein's Colonel Manning.

Science fiction makes strange bedfellows. Throughout his career, Heinlein was a source of outrage to those of liberal sensibilities, having written novels that can be read as endorsing McCarthyism (*The Puppet Masters*, 1951), genocidal warfare (*Starship Troopers*, 1959), and racist paranoia (*Farnham's Freehold*, 1964). He even anticipated, in *The Moon Is a Harsh Mistress* (1964), the latest delusional fancy of the lunatic right fringe: the notion that the United Nations is about to establish dominion over the forces of freedom, whose last hope is to band together as armed militias. Despite this track record, Heinlein was a favorite writer of the '60s counterculture, and not simply for the New Age fantasies of *Stranger in a Strange Land* (1961), but because he is a prime example of Marcuse's theory of "repressive desublimation."[11] In other words, he knew what his primary audience of adolescent and college-age male readers was in the market for, and he spiced his right-wing politics with as much overt sexuality as the traffic would bear. In this he was in sync with such other popular authors of the Cold War era as Mickey Spillane (whose *Kiss Me Deadly* became, in its movie version, one of the great monuments of apocalyptic daydreaming) and Ian Fleming (whose James Bond novels were propelled to the top of the best-seller lists by President Kennedy's endorsement). In all but his most flagrantly fascistic fantasies, Heinlein was able to transcend conventional dichotomies of right and left, because (1) readers were less sophisticated about the coding of political messages into "entertainment," (2) "serious" readers didn't bother with SF, and so any controversy that Heinlein might stir among leftish SF readers never reverberated beyond the SF frogpond, and (3) Heinlein's novels were irresistibly readable.

The Puppet Masters, Heinlein's most inspired evocation of Cold War terror, is a crafty blend of love and rockets, of winking sexual titillation and gung-ho violence that subdues nuclear dread by sheer bluster and divine right. Heinlein solves the moral conundrum of Hiroshima (Is it okay to kill millions of innocent civilians?) by positing extraterrestrial enemies of exemplary vileness. The aliens of *The Puppet Masters* are parasitical slugs who plug into the human nervous system and use their hosts as slaves, whose first task is to help turn the entire human race into the aliens' "puppets," much as in the derivative *Invasion of the Body Snatchers* (filmed 1956). Unless suspects are stripped naked, it's impossible to tell if they are hosting an alien or are still human. The policeman on the beat, your doctor, your spouse: any of them might be an alien. Just so, according to McCarthyite witch-hunters of that era, Commie spies were everywhere, having disguised themselves as New Deal liberals. There is only one way to deal with such an enemy, according to Heinlein's narrator: total annihilation. The story ends with an exhortation that shows atomic age saber-rattling at its most rhapsodic:

> I am . . . a combat trooper, as is every one of us, from chaplain to cook. This is for keeps and we intend to show those slugs that they made the mistake of tangling with the toughest, meanest, deadliest, most unrelenting—and ablest—form of life in this section of space, a critter that can be killed but can't be tamed. . . .
>
> Whether we make it or not, the human race has got to keep up its well-earned reputation for ferocity. The price of freedom is the willingness to do sudden battle, anywhere, anytime, and with utter recklessness. If we did not learn that from the slugs, well—"Dinosaurs, move over! We are ready to become extinct!" . . .
>
> We are about to transship. I feel exhilarated. Puppet masters—the free men are coming to kill you!
>
> *Death and Destruction!*[12]

Reading *The Puppet Masters* at age twelve, I did not think to decipher subtexts that now seem to me as visible as sore thumbs. I was uninformed and unconcerned. Current Events was something we did at school for fifteen minutes on Friday. I delivered newspapers but didn't read them. My mother listened to Arthur Godfrey and soap operas but paid scant attention to news programs. Television had not yet reached Fairmont, Minnesota, and even when my family moved to the Twin Cities, where TV could digest them, news programming had not yet reached that level of showmanship where it might command the attention of teenagers.

So my take on *The Puppet Masters* was that of most of its other early readers: I was terrified by the slugs, by the idea of being controlled by an alien Something that hooked into the base of my spine and made me do terrible things. When Heinlein's narrator describes his experience as the host of a parasite, I could identify:

> I saw things around me with a curious double vision, as if I stared through rippling water—yet I felt no surprise and no curiosity. I moved like a sleepwalker, unaware of what I was about to do—but I was wide awake. . . .
>
> I felt no emotion most of the time, except the contentment that comes from work which needs to be done. That was on the conscious level; someplace, more levels down than I understand about, I was excruciatingly unhappy, terrified, and filled with guilt, but that was down, 'way down, locked, suppressed; I was hardly aware of it and not affected by it.[13]

This "double vision" could pass muster very well as a description for what has also been called "bad faith" or "false consciousness." It is surely the condition of Winston Smith's soul in the last pages of *1984*, when he has learned to love Big Brother. In this respect, Heinlein's equating of his aliens with the Russians is not a gratuitous slur. Heinlein drives the point home in a key passage: "I wondered why the [slugs] had not attacked Russia first: the place seemed tailor-made for them. On second thought, I wondered if they had. On third thought, I wondered what difference it would make."[14]

It is the third thought that should give us pause, because if it is not just a passing cleverness, it must be understood to mean that Russia should be confronted with just the same "ferocity," the same "willingness to do sudden battle," the same "utter recklessness" that is required to combat the slugs. "Better Dead Than Red" was then a favored slogan of those who favored a preemptive nuclear strike against the Communists. Heinlein seems to be saying the same thing.

Paul Boyer sees an emblematic significance in James Cagney's death scene at the end of *White Heat* (1949), as the gangster, trapped on top of a gigantic spherical gas tank, chooses to self-destruct with a cry, "Top of the world, Ma!" To Boyer this is equivalent to the nuclear suicide that psychiatrist Franz Alexander had forebodings of in his 1949 essay, "The Bomb and the Human Psyche": "Having contrived the means of his destruction . . . man might find irresistible the temptation to escape forever the stresses of the atomic age and subconsciously conclude that 'a painless end' was preferable to 'endless pain.'"[15]

Perhaps what is scariest about Heinlein's bellicose fantasies is their death-wish component. His aggression may well be the bluster of a teenage boy who affects a skull-and-crossbones ear stud (like the young hero of *Starship Troopers*) in order to command the respect of, at the very least, his bathroom mirror. What may well be more dangerous in the long term are those feelings of excruciating unhappiness, terror, and guilt, all stuffed "down, 'way down, locked, suppressed."

Robert Heinlein was born in 1907. A graduate of the U.S. Naval Academy at Annapolis, he had to retire from active service in 1934 when he contracted tuberculosis. He began publishing SF in 1939, and in one of his first stories, he predicted the use of nuclear bombs to bring World War II to an end. When Hiroshima was bombed, Heinlein was thirty-eight and his character set in concrete. He may have felt a certain tingle of terror or guilt at an unconscious level, but his likely response to the event would be a simple, smug, "I told you so."

Philip Dick, our second laureate of nuclear dread (and excitement), was born in 1928. He came of age, quite literally, *with* the bomb. According to a boyhood friend, Leon Rimov, "he was really, really obsessed" with news of the war, charting its progress with maps and drawings.[16] When the bomb fell on Hiroshima, he was still sixteen, and the event sent shock waves through the rest of his life. When he began to publish

SF in the early '50s, his stories and novels were set, almost as a matter of course, in a post-nuclear-holocaust future. As early as 1955, in an essay published in an SF fanzine, Dick set down the theory that would govern his practice in dozens of stories and novels yet to be written:

> All responsible writers, to some degree, have become involuntary criers of doom, because doom is in the wind. . . . In science fiction a writer is not merely inclined to act out the Cassandra role he is absolutely obliged to. . . .
>
> [But since] doom stories become monotonous . . . perhaps we should take the doom for granted and go on from there.
>
> Make the ruined world of ash a premise: state it in paragraph one and get it over with, rather than winding up with it at the very end. And make the central theme of the story an attempt by the characters to solve the problem of postwar survival.[17]

And so, in a long succession of stories beginning with "The Variable Man," "Second Variety," "The Defenders," and "Foster You're Dead" (all four from 1953), and in a goodly number of his novels (including *The World Jones Made, Time Out of Joint, The Man in the High Castle, The Game Players of Titan, Dr. Bloodmoney, The Zap Gun, Do Androids Dream of Electric Sheep?*), Dick plays endless variations on the theme of Cold War terror and the arms race.

In some cases, nuclear dread is not the focus of the plot but serves simply as the mise-en-scène. Indeed, one can't really characterize Dick's take on these matters as one of dread. As in the subtitle of Kubrick's epochal 1962 film, *Dr. Strangelove (or, How I Learned to Stop Worrying and Love the Bomb)*, Dick had supped with horrors on such a regular basis that many of his most memorable apocalyptic tales are comic, lighthearted, even pastoral. Okay, he seems to be saying, so we're all going to be incinerated the day after tomorrow. That's unfortunate. But, *meanwhile*, have you heard the one about the Martian and the Whore of Babylon? Dick's humor, like that of the two other Cold War writers with whom he has the closest affinity, Joseph Heller *(Catch 22)* and Kurt Vonnegut *(Slaughterhouse Five)*, is inseparable from his sense of imminent disaster.

Dick's first novel to treat the nuclear theme with characteristically mordant wit was *Time Out of Joint* in 1959. The novel's hero, Ragle Gumm, is, like so many other Dick heros, a metaphorical self-portrait. Gumm earns his living not by writing SF but by a still odder task: each day he solves the "Where-Will-the-Little-Green-Man-Be-Next?" contest in the daily paper. In fact, Gumm (who thinks he is living in 1958) is actually using his psychic gifts to chart the American defense strategy in the year 1998. The military has built a ramshackle World of 1958 just for him, so that he can make strategic decisions without being distressed by the nuclear sword of Damocles poised over his head. Most Dick enthusiasts have relished the Potemkin village of *Time Out of Joint* without noting the justice of its moral vision, which can be summed up quite simply: everyone who is reading the news *is responsible for the news*. Or, put another way, the nuclear sword of Damocles that hangs over our heads is paid for by our taxes and sustained by our tacit permission.

There is a moment at the end of one of the great SF novels of the '50s, Alfred Bester's *The Stars My Destination*, that prefigures the moral gestalt of *Time Out of Joint*. In Bester's novel, after a great many thrilling adventures, the hero, Gully Foyle, discovers the nature of the McGuffin[18] he and the rest of the cast have been pursuing through the book. It's called PyrE, and "it's a thermonuclear explosive that's detonated by . . . the desire of anyone to detonate, directed at it. That brings it to critical mass if it's not insulated by Inert Lead Isotope." Gully Foyle's response to this revelation is to discover the secret of PyrE to the entire world, as he "jauntes," telekinetically, from city to city:

> Foyle withdrew a slug of PyrE, the color of iodine crystals, the size of a cigarette . . . one pound of transplutonian isotope in solid solution.
>
> "PyrE!" he roared to the mob. "Take it! Keep it! It's your future. PyrE!"
>
> [He jauntes to San Francisco, to Nome, delivering the same message. Then he returns to confront the members of the power elite from whom he stole this Promethean fire, and to return the PyrE that still remains.]
>
> "There's enough left for a war. Plenty for destruction . . . annihilation . . . if you dare." He was laughing and sobbing, in hysterical triumph. "Millions for defense, but not one cent for survival."[19]

For many years, SF readers of a literary bent considered *The Stars My Destination* the best SF novel ever written, an enthusiasm that had its source, to some degree, in the cathartic force of Bester's PyrE-technical finale. PyrE is an even better metaphor for nuclear Liebestod than the suitcase bomb that explodes at the end of the movie version of *Kiss Me Deadly*. The simple message of PyrE is: You want Armageddon? Here, you got it.

In one way that is a liberating message; in another way it's balderdash because, in reality, the people who might be killed by the bomb have no voice in its deployment. From a pragmatic point of view, history is something that happens to us, which is why so many people choose to avert their gaze. But not Dick. In his next great novel, *The Man in the High Castle*, Dick jettisons the comforting notion of an average man whose psychic powers allow him to be the fulcrum of world history and imagines a world much more like our own, governed by juggernaut political forces. Yet this more "realistic" world of A.D. 1962 is also marvelously unlike the 1962 in which it was published, for the Axis powers have won World War II, and the United States has been divided, like postwar Germany, into zones of occupation. Germany rules the eastern seaboard and some midwestern states, while Japan controls the West Coast. A triumphant Germany has depopulated the former USSR and Africa, drained the Mediterranean to create new farmlands, and instituted concentration camps throughout North America. The Germans have also sent rockets to the moon, Mars, and Venus. On the West Coast, the Japanese occupation forces have instituted a regime that ironically mirrors the American occupation of Japan. The conquerors have become discerning collectors of Americana—everything from Civil War memorabilia to Mickey Mouse wristwatches—while the conquered are busy assimilating Japanese culture.

Meanwhile, as they consolidate their triumph, the Germans and Japanese are engaged in their own cold war. The Germans dispatch U.S. troops to invade Japanese puppet states in South America. Dick even gives a shrewd estimate on how the space race factors in to such a contest, as in this bit of dialogue between two Japanese bureaucrats:

> "The Home Islands [Japan] take the view that Germany's scheme to reduce the populations of Europe and Northern Asia to the status of

> slaves—plus murdering all intellectuals, bourgeois elements, patriotic youth and what not—has been an economic catastrophe. Only the formidable technological achievements of German science and industry have saved them. Miracle weapons, so to speak."
>
> "Yes. . . . As did their miracle weapons V-one and V-two and their jet fighters in the war."
>
> "It is a sleight-of-hand business. Mainly, their uses of atomic energy have kept things together. And the diversion of their circuslike rocket travel to Mars and Venus. . . . For all their thrilling import, such traffic [has] yielded nothing of economic worth. . . . Most high-placed Nazis are refusing to face facts vis-à-vis their economic plight."

Better than any other SF writer of his time, Dick understood that science fiction is not about predicting the future but examining the present. The world that Aldous Huxley depicts in *Brave New World* is not some hypothetical tomorrow; it is the thrilled and horrified reaction of an upper-class Englishman reacting to his first visit to Jazz Age America, a nation that values the "feelies" (Hollywood movies) above the tragic vision of Shakespeare. The world that George Orwell envisaged in *1984* is simply a darker version of the year it was written, 1948. To appreciate such satires fully, one must exercise a kind of double vision, savoring the wilder flights of fancy but aware, all the while, of the authors' direct hits on contemporary targets.

"Serious" readers tend to deprecate such strategies, perhaps because they have an inkling that they are being lampooned along with their enemies. Early leftist readers of *1984* and *Animal Farm* were hostile to those books because they thought Orwell was attacking socialism (as he was), while readers of the right applauded, unaware that his description of the Ministry of Truth cut both ways. Huxley's vision of a totalitarian state based on hedonism deflects (or invites) criticism in much the same way. If he had simply ranted against the values of Hollywood—glamour, sensation, excitement, *skin*—he would have come across as another sourpuss like the Pope or Solzhenitsyn. The trick of great satire is to body forth a world in which the values one reprehends are knitted into the fabric of daily life, a world in which the consciousness of every character has been transformed by the nightmare it inhabits.

This is what Dick does so resonantly in *The Man in the High Castle.* As the point of view shifts from that of one character to another, we see how the *faits accomplis* of history determine the thoughts it is possible to think. A San Francisco antique dealer toadies up to his Japanese customers with a servility that is at once timeless and specific to the new Zeitgeist. An intelligent Italian hit man discusses world affairs with the blinkered but keen perspective of a proletarian Machiavelli. Every line of dialogue is shaped by the world Dick has imagined—as are our own conversations, could we but overhear them from a sufficient distance. The alternate world of Dick's creation allows that distance.

It also allows a needful deniability. Dick's imagination was fueled by a paranoia that was sometimes merely playful and sometimes in dead earnest. As an SF writer, unheeded because unread, he enjoyed liberties denied to other writers of the McCarthy era, when his first works were published. He was not alone in enjoying such advantages. Science fiction continues to be a forum in which the heterodox opinions of both the left and right may be expressed with relative impunity. Who cares? It's just trash anyhow.[20]

Dick blossomed as a writer after finishing *The Man in the High Castle* in 1961, producing in rapid succession a string of novels that was like a barrage of fireworks at the climax of a Fourth of July celebration: *Martian Time-Slip* (1963), *Dr. Bloodmoney, or How We Got Along After the Bomb* (1963), *The Game Players of Titan* (1963), *The Simulacra* (1964), *Clans of the Alphane Moon* (1964), *The Penultimate Truth* (1964), *The Three Stigmata of Palmer Eldritch* (1965), and *The Zap Gun* (1966). Composers more often than novelists enjoy such a heyday of do-it-again successes. In Dick's case, however, they were successes only of esteem—and that only within the small part of the SF readership tuned to his wavelength.

Of those eight novels, *The Penultimate Truth*, while far from the best, represents his final statement on the nuclear arms race, which is simply this: ignore it; it doesn't exist. In *The Penultimate Truth,* the United States has become a literally stratified society, with "tankers" living in underground bunkers as a refuge from existing fallout and the threat of further nuclear war, while an upper class of "Yance-men"—politicians and media bureaucrats—enjoys an ideal exurban existence upstairs, where, instead of the devastation viewable on the tankers' TVs, all is green and lush and pleasantly exclusive. There was no war and is no threat. Just as some cranks maintain that the *Apollo* moon landings were all an imposture, so in this novel the entire Cold War and arms race are scripted inventions—"news stories" constantly being rewritten by the minions of an Orwellian Ministry of Truth. And what is the purpose of this enormous imposture? One of the Yance-men (scriptwriters for the nation's simulated "spir-pol-mil leader," Talbot Yancy) confides it to the reader in an internal monologue. That purpose is

> for each of us [Yance-men] here to augment our retinues . . . which wait on us, follow us, dig for us, build, scrape and bow . . . you've made us barons in baronial castles, and you are the Nibelungen, the dwarves in the mines; you labor for us. And we give you back—reading matter.[21]

This may not be the literal truth of the world of 1964, but it was then, and remains, a liberating illusion. Like Voltaire's God, if it did not exist, we would have to invent it—and Dick did. Indeed, he did it repeatedly, in a musicianly way, with theme and variation. First, in *Time Out of Joint*, he posited an unreal 1958, a seemingly quotidian suburban life that is actually the war room of an unseen nuclear conflict. In *The Man in the High Castle*, we are offered an alternative history of the world, in which the title character is a science-fiction writer who has imagined another alternate world (our own, or very nearly) in which the Allies won World War II, and this alternate world (it is suggested toward the end of the novel) is the *real* world. Finally, in *The Penultimate Truth*, Dick declares that life—the life we were compelled to live in 1963—is a dream.

And, in effect, he was right. The bombs didn't fall—not during the Bay of Pigs Crisis, not after Kennedy's assassination, not through the entire Vietnam War. If one had simply ignored the arms race through that entire period and on to the present day, life would have gone on, and we could have avoided our nuclear worries by a simple act of denial. And that is just what we did. Here is how the historian Paul Boyer describes the period that followed the heyday of high nuclear anxiety from 1959 *(On the Beach)* through 1963 *(Dr. Strangelove)*:

> What one does see after 1963, however, as in 1947–54, is a sharp decline in culturally expressed engagement with the issue [of nuclear peril]. With apologies to Raymond Chandler, one might call this the Era of the Big Sleep. Public opinion data reflect the shift. In 1959, 64 percent of Americans listed nuclear war as the nation's most urgent problem. By 1964, the figure had dropped to 16 percent. Soon it vanished entirely from the surveys. An early 1970s study of the treatment of the nuclear arms race in American education journals found the subject almost totally ignored. "The atom bomb is a dead issue," concluded a sociologist studying student attitudes in 1973.[22]

Boyer offers an assortment of possible explanations for this sudden swerve of the Zeitgeist: that the test ban treaty of 1963, by putting an end to atmospheric testing, offered an illusion of diminished risk (out of sight is out of mind); that nuclear power was getting good publicity from the first nuclear power plants then coming into use; and, chiefly, that "in the later 60s the Vietnam War absorbed nearly every available drop of antiwar energy. . . . As the source of powerful television images, the war and the domestic turmoil it engendered had an immediacy the more abstract nuclear weapons issue could not begin to match."[23]

I would offer two other reasons besides those adduced by Boyer. The first is that the 1964 presidential race between Lyndon Johnson and Barry Goldwater became a public referendum on the issue of nuclear peril, thanks in large part to the TV ad produced by Doyle Dane Bernbach, which showed a nuclear mushroom as the likely result of a Goldwater presidency. I was working as a lower-echelon Yance-man at DDB at that time, but it is not company loyalty that makes me think that ad was as crucial to Johnson's election as, later, the Willy Horton ads would be to the election of President Bush. Many political commentators have cited that ad campaign as the moment when national politics began to be dominated (as per *The Penultimate Truth*) by the media. In effect, voters in 1964 were not so much voting *for* Johnson as *against* the bomb. When they won the election by a decisive 60-40 margin, they could sleep more easily: the Bomb had been defeated.

The other reason was the release in 1963 of Stanley Kubrick's *Dr. Strangelove (or, How I Learned to Stop Worrying and Love the Bomb)*, a black comedy that acted like a depth charge on the American collective unconscious. Based on a run-of-the-mill techno-thriller, *Red Alert,* by Peter George, Kubrick's movie, which was scripted by himself, George, and humorist Terry Southern, had the temerity to laugh at the prospect of nuclear Armageddon and everything connected to it: lunatic-fringe right-wing generals, conscript Nazi rocket engineers, and crossed psychic circuitries that equated the bomb with sex and/or contact sports. From the perspective of a third of a century later, the movie's humor can sometimes seem sophomoric (as in the names that characters were given: the mad general is Jack D. Ripper; the Soviet premier is Dmitri Kissoff), sometimes hysterical, in a pathological sense (as when Peter Sellers, as a caricatured Werner von Braun, can't control his impulse to give a Nazi salute), but for those who saw it when it came out, it was riotously, convulsingly, liberatingly hilarious. It was catharsis by comedy. The stored-up anxieties of almost two decades of nuclear anxiety were released in a saturnalia of laughter that for many of us actually made good on the promise of the subtitle: we learned to stop worrying and to live with (if not to love) the bomb.

Dick was one of those directly influenced. Although the title was suggested by his editor at Ace Books, Donald Wolheim (who was responsible for the first publication of sixteen of Dick's novels), and not by Dick himself, *Dr. Bloodmoney or How We Got Along After the Bomb* is clearly a product of the post-Strangelove era. The title character, Dr. Bruno Bluthgeld (German for "bloodmoney"), is based not on Wernher von Braun but on the scarier (because more influential) Edward Teller, the Father of the H-Bomb. There are many dire events as the plot unfolds, but the tone of the book is pastoral. Its postnuclear landscapes are populated with mutant dogs that have learned the rudiments of speech and evolved rats that play nose flutes. In effect, Dick is anticipating the feckless wisdom of *The Penultimate Truth*, written a year later, by treating the Bomb as a kind of global magic potion that can shift our world into a Wonderland embellished with strange beasts and allegorical mutants.

Many SF writers took Dick's hint and produced postnuclear playgrounds of a similarly allegorical-decorative nature: Harlan Ellison in his 1969 novella, "A Boy and His Dog" (filmed in 1976); Roger Zelazny in *Damnation Alley* (1969, filmed in 1977), with its mutated, carnivorous, high-camp cockroaches; and Steve Wilson in the best of all post-Apocalyptic biker fantasies, *The Lost Traveller* (1976), in which the Los Angeles chapter of Hell's Angels rules supreme following World War III. Though not filmed, Wilson's book was obviously the inspiration for *Mad Max 2: The Road Warrior* of 1979.

Those books and movies (and many more besides) represent the decadence of nuclear dread, the baroque nightmares of the era Boyer calls The Big Sleep. One may reprehend them for their essential lack of moral earnestness and novelistic verisimilitude or applaud them for the way they have quieted nuclear dread by turning it into the tropes of a new gothicism. Applause has been the general rule, for the all-sufficient reason that we need our nightmares to be somehow tolerable. Poe knew this; it was the secret of his success. He opened the Pandora's box of his Id and played with the toys he found there. His successors, in the fields of both mystery and science fiction, have needed no other instruction. Even the strangest love of all, the death wish, can be taught to sing enchantingly.

That is one way to account for the sea change in our apprehension of the nuclear age we find ourselves (still) alive in. But there is also a technological component to that transaction. Another and more potent technology than that of the Bomb came to dominate our lives and culture in the years we have been considering: the technology that was becoming our second nature all through the era of nuclear dread and that rules us now: television.

Chapter 5

STAR TREK, OR THE FUTURE AS A LIFESTYLE

Television and science fiction, though they've lived a long time together, have not enjoyed a happy relationship. Television, the dominant and more affluent partner, has been blithely unaware of this fact, which is often the way in such cases. Science fiction, on the other hand, has felt, well . . . used.

Here is how Peter Nicholls, coeditor of the generally even-toned *Encyclopedia of Science Fiction,* states the case in the 1993 edition of that reference work in an article under the heading "Television":

> The pressures towards conformity and formula, especially in the USA but also in the UK, have meant that televised sf, in a history spanning over 40 years, has never approached the intellectual excitement of the best written sf, or indeed the best sf in the cinema. Because televised sf cleaves to the expected, we are seldom surprised by it: we seldom feel any sense of wonder or even stimulation. At best we are amused by the occasional adroit variation on a familiar theme, or by bits of rather good acting. Televised sf is a cultural scandal; it is, on the average, so much worse than it could be or needs to be. But there seems to be no way to combat the entropic forces that make it that way. The tv industry is something of a "closed shop", with its own well-established

> writers and producers—one reason why it has generally proven inhospitable to sf writers—and it is difficult to influence from the outside. Until this is done, the standard of televised sf will not improve.[1]

Many SF writers would concur with Nicholls—especially the few who've occasionally enjoyed Writers' Guild wages (an hour-long screenplay for a widely syndicated show earns a mid-rank writer more money than a full-length novel). But Nicholls's base is in academia (albeit an academia that predates the leveling relativism of cultural studies departments). Whatever sourness he may express derives from his frustrations as a consumer rather than a producer. It's a fact: the SF on TV (movie reruns excepted) ranges from rather to very dumb.

A brief history. *Captain Video* premiered in 1949, a low-budgeted daily serial aimed at preteens, broadcast live, with cheesy special effects. Its success spawned imitators: *Buck Rogers* (1950–1951), *Tom Corbett, Space Cadet* (1950–1955), *Space Patrol* (1950–1955), and others—all of them pure kiddy-vid, and so primitive technically as to be beyond all hope of syndication nowadays. Modern children, even in Nepal, simply demand more in the way of FX. These series fizzled out in the '50s, followed by two moderately successful anthology series, *The Twilight Zone* (1959–1964) and *Outer Limits* (1963–1965), and many more one- or two-season fizzles. These anthology programs more often featured fantasy plots, or stories with borderline SF elements, such as *Twilight Zone* adaptations of Jerome Bixby's "It's a Good Life," a tale of an omnipotent boy who repeats his birthday forever, or Ray Bradbury's sentimental "I Sing the Body Electric," about a robot grandmother. Robots and psychic children lent themselves better than rocket ships to the budgetary limitations of '60s TV.

But then came *Star Trek*. At first *Star Trek* seemed doomed to follow the pattern of earlier SF TV series, fizzling out after three seasons (1966–1969). Peter Nicholls's evaluation of those first seventy-nine shows is uncompromisingly dour:

> For fans of written sf, *Star Trek* can seldom have seemed challenging in any way, as it rarely departed from sf stereotypes. . . . The space opera format was not used with any great imagination. A typical

> episode would face the crew with Alien superbeings. . . . The formula seldom varied. Many adult viewers came to feel that the series was bland, repetitious, scientifically mediocre and, in its earnest moralizing, trite. The effort to please all and offend none was evident in the inclusion of a token Russian, a token Asiatic and, together in the person of actress Nichelle Nichols, a token Black and token woman. The defect in this liberal internationalism was that all these characters behaved in a traditional Anglo-Saxon Protestant manner; only Spock was a truly original creation.[2]

Nicholls's bile is surely provoked not just by those elements of *Star Trek* that are, indeed, meretricious but by the series' subsequent, virtually all-conquering success. After its first cancellation, *Star Trek* continued to be shown in syndication, gathering ever more fans and becoming a cult phenomenon, engendering a spate of paperback novelizations and spinoffs, a TV cartoon series, and, in the course of time, high-budget feature films and, at last, three further *Star Trek* TV series with new casts. New *Star Trek* books pop up regularly on best-seller lists, and (unlike most nonfranchise fiction) they remain in print. The Barnes & Noble store nearest where I live has forty feet of shelf space devoted to *Star Trek* books, a virtual lifetime of reading. Such written-for-hire novels became a kind of sweatshop for SF writers who'd fallen on hard times, beginning with James Blish (1921–1975), who supported himself during his last years of debility by novelizing eleven *Star Trek* TV scripts. It is a common fate in an age of franchises, as those who'd once been the proud owners of their own independent diners find themselves earning the minimum wage flipping essentially the same burgers for McDonald's. Little wonder then that within the SF community, *Star Trek* is regarded with all the fondness Ben-Hur felt for the Roman army.

Any very large success, especially one that becomes an institution, bears examining. Putting aside any personal distaste, we must ask: Why *Star Trek?* If, as Nicholls grumbles, it is true that the show is seldom "challenging," that could be because few audiences like to be challenged in the sense Nicholls intends. A "challenge," after all, is traditionally the prelude to a duel, not to a half-hour of light entertainment. Any artist's first order of business is not to challenge but to entice.

As to "bland" and "repetitious," why not? Blandness and repetition can be comforting, and comfort is a major desideratum in bedtime stories. *Star Trek*'s target audience, after all, is ages nine to thirteen. Though perhaps less skilled in reading prose fiction than children of the '40s and '50s, they have more savvy about media. The special effects and acting skills in evidence on *Star Trek* are several cuts above those of earlier TV series like *Buck Rogers*. Improved production values have their own affective significance. As Santa knows, children are conscious of how much money is spent on them. On the other hand, children tend not to be venturesome consumers, happiest when they can eat the same few favorite foods. One major component of *Star Trek*'s success is that it's pizza for lunch, every day of the week.

TV is a visual medium. This might seem too obvious to mention, but it's something that writers, when they assess the content of a TV program, seldom bear in mind. If you want to *see* how a TV series works, watch one installment with the sound off. I learned this valuable lesson from my Aunt Aurelia, who explained that the special appeal of the soap operas she watched every day in her retirement years was not the complicated plots (which she couldn't always keep track of) but the way all the characters were so well dressed and lived in such tastefully decorated homes. She watched the soaps as once people went to the opera—not for the sake of the story, which was often preposterous or unseemly, but to see what people were wearing.

Television's forte has always been its menu of role models and codes of etiquette suitable for a wide range of viewers. Male authority figures—anchormen, lawyers, the owners of basketball teams—must wear suits, as do the more upscale criminals when they appear in noir movies or as defendants at murder trials. Other modes of male attire also have their distinctive TV genres to serve as fashion runways. The western, most popular in the late '50s and through the '60s, ushered in the Age of Denim. What had been the livery of menial labor became the uniform of youth and an all-purpose form of mufti for city dwellers who would be both egalitarian and low profile. In the '70s the uniforms of the military and of professional sports teams became common on the streets in proportion as they were visible on the small screen.

But how is one to dress in the next century, which will be arriving soon at neighborhood theaters? Not in a suit, those symbols of the black-and-white past, and not in the jeans one wears to school. Surely the future offers something spiffier than that! Before *Star Trek*, the SF in the movies, comics, and TV offered two possibilities: either one dressed like real astronauts, in cumbersome spacesuits; or, like characters in *Buck Rogers*, one's wardrobe conflated all periods of history, as though the future were to be one big Beaux Arts ball with Viking helmets and brass brassieres, Renaissance doublets with flaring collars, capes, sashes, body tights—whatever, indeed, could easily be rented from a costume supplier. The trouble with both these approaches was that they could not be merchandised. The franchising possibilities for blimpy space suits with bubble helmets are limited; ditto the mass market prospects for the Emperor Ming look.

Star Trek's solution to this dilemma was pajamas: a collarless, capeless, looser-fitting Superman suit that can't be mistaken for any known form of streetwear of the late '60s. The pajamas varied in color but were otherwise one-style-fits-all, with only modest pins and badges to distinguish rank. While sports uniforms are blazoned with numbers and names, and a man in a gray suit may yet flaunt a bright tie, in *Star Trek* pajamas all men are equal from the neck down. The lesson such a uniform teaches is that conformity will be the order of the day in the future even more than in the present. And at age twelve, the urge to conform—and therefore to *belong*—can amount to a passion.[3]

And where will this final harmony obtain? Turn down the sound, and look at the show's sets. Where in real life would one be likeliest to encounter an environment so brightly and blankly geometric and uniformly lighted? Dress everyone in suits instead of pajamas, and it's clear that the Starship *Enterprise* is actually an office disguised as the Future. What other future, after all, is a likelier destination for most of the younger viewers who will graduate to the *Enterprise* from schoolrooms that are also visual analogs to the show's sets?

Finally, with respect to its formulaic plots, in which the little beehive of the *Enterprise* confronts, each week, some new variety of non-pajama-wearing misfits or aliens, *Star Trek* is offering its viewers essentially the same parables of success-through-team-effort that can be found on such later workplace-centered sitcoms as *The Mary Tyler Moore Show* and *Designing Women*. And in this respect, *Star Trek* did boldly go where no such TV series had gone before in being among the first TV venues to show how to behave in an environment—the office or the schoolroom—where differences of gender and race have been declared officially invisible. Before political correctness or multiculturalism had become debating topics, *Star Trek* was their prophet.

Nor can it be said that this prophet has gone unrecognized in his or her own country. For those Trekkies who might miss the series' secret wisdom, Morrow has brought out an "unauthorized" self-help book, *Boldly Live As You've Never Lived Before* by Richard Raben and Hiyaguha Cohen, which (according to its flap copy) "enables you to recognize your own heroic qualities and unleash their tremendous power." Raben has been "the director of training and quality management for large companies," while Cohen "has an active career-counseling and corporate-outplacement practice." As befits their backgrounds, the authors find *Star Trek* to be full of lessons in personnel management. The first lesson is to know your heroic character type and its job description. There are four choices: the Worf-like *Warriors* (who are commended to careers as sales reps and policemen); the Spock-like *Analysts* (accountantcy and software engineering for them); the *Relaters*, like Troi or Kes (teachers, nurses, personnel directors); or the strong *Leaders*, like Kirk or Picard, who are destined for the top of every pyramid of command and organization tree.

Now suppose you've decided you're the leading-man type. Here's what Raben and Cohen suggest you do. First, study the experts in their best *Star Trek* episodes (cited in the book). Then "*Maintain direct eye contact*, but don't glare like a Warrior or flirt like a Relater. The Leader uses eye contact like a handshake: to open a channel for businesslike communications." Then, "*Dress for success*. Leaders look smart. Would you want to follow somebody who had shirttails sticking out, crumpled clothes, or ring-around-the-collar? You must look fit to lead."[4] Relaters, by contrast, are advised to be touchy-feely, smile and hug a lot, and "use soft, affectionate eye-contact." At the same time: "Show that you accept the other person's thoughts and feelings without judging. Don't offer advice; give support only. You can say things like: 'It must be very hard for you right now' or 'I don't blame you for feeling that way' or 'How can I better understand what you're going through?'"[5]

The rewards promised to those who emulate their *Star Trek* avatars are nothing less than utopian:

> Imagine how great life would be if everyone recognized your unique genius, and if you never envied anyone else's. Can you conceive of a work environment where every single person on staff appreciated, respected, and understood what you do? Where all of your colleagues supported and encouraged you? One where you got assigned to roles that maximized your talents, where your boss trusted you more than you trust yourself, and where power struggles did not exist because everyone loved his or her own job?
>
> That's life on a *Star Trek* vessel. . . .
>
> . . . In contrast, what happens in the corporate world if some upstart underling gets promoted? Bloodshed! People resign in a huff, try to sabotage the new boss, gossip wickedly. . . .
>
> The *Star Trek* crew can't afford such trouble. If jealousy reigns on the Bridge and people fight for power, they certainly won't survive many missions. Can you imagine facing a Romulan warship while fighting over who gets to fire the phasers?[6]

In Raben and Cohen's accounting, the *Enterprise* is nothing less than a utopia, an ideal social environment to be given serious consideration as a blueprint for a future America that would be one smoothly running military operation without conflict between rich and poor, male and female, black and white. These ideals weren't preached by the show's scriptwriters; they were shown as a fait accompli, and one vividly at odds with the real circumstances of their era, when protest against the Vietnam War was at its height, when Martin Luther King, Jr., had just been murdered, when feminists had banded together to form NOW and press for the ill-fated equal rights amendment.

There has always been an overlap between SF and utopian literature. Early utopias, like those by Thomas More and Tommaso Campanella, were located in remote corners of the globe rather than on other planets or in the far future, and they took a form closer to the essay than to narrative fiction. But certain distinctive traits are continuous from More's *Utopia* to the latest cyberpunk paperback. One of those is the utopian obsession with good grooming. In the Good Place (as one may translate the Greek roots *eu-topos*, though it may also be *ou-topos*, "nowhere") it is important that one look *comme il faut*. As on the *Enterprise*, simple clothing has been the general utopian rule. In More's *Utopia* of 1515 and in Campanella's *City of the Sun* of 1602, deviations from strict sumptuary laws are punishable by law, often severely. The utopian impulse makes school principals of us all, from Plato to Ursula Le Guin, fanatical to enforce Savonarolean dress codes and to legislate good manners. The first world-class American utopia, Edward Bellamy's *Looking Backward* of 1888, foresees the year 2000 as a paradise of political correctness, wherein "no officer is so high that he would dare display an overbearing manner toward a workman of the lowest class. As for churlishness or rudeness by an official of any sort, in his relations to the public, not one among minor offenses is more sure of a prompt penalty than this. Not only justice but civility is enforced by our judges in all sorts of intercourse."[7]

Bellamy's book is the first sustained fictional presentation of the modern welfare state, in which "no man any more has any care for the morrow, either for himself or his children, for the nation guarantees the nurture, education, and comfortable maintenance of every citizen from the cradle to the grave." The idea was hugely popular in its day, and Bellamy Clubs were formed by middle-class enthusiasts who aspired to become the administrators of a Bellamist system. Introducing *Looking Backward* in a 1960 paperback edition, Erich Fromm buys everything Bellamy is selling, praising this earlier version of the *Starship Enterprise*, where, as Fromm explains, "There is no individual antagonism, but a sense of solidarity and love. . . . They are frank, and they do not lie, and there is complete equality of the sexes, with no need for deceit or manipulation. In other words, it is a society in which the religion of brotherly love and solidarity has been realized."[8] It is also one big gulag where everyone between ages twenty-five and forty-five is conscripted into an industrial army employed in producing an abundance of modern conveniences. In short, Bellamy provided the theory for the world we would be living in now—if only its residents would behave as the theory requires.

Science-fiction writers have generally steered clear of writing out-and-out utopias from a sense that they are likely to be preachy, undramatic, and, like Bellamy's, terminally genteel. Good people leading wholesome lives in conflict-free polities are not the stuff novels are made of. (Dystopias, on the other hand, are an SF staple, and when non-SF writers venture into the future tense, their works usually take a dystopian form. Some have become classics: Huxley's *Brave New World*, Orwell's *1984*, Atwood's *The Handmaid's Tale*. Others were literary belly flops: Paul Theroux's *O-Zone*, P. D. James's *The Children of Men*, and Doris Lessing's *Canopus in Argos: Archives* series. Whatever its literary merit, each dystopia, like Tolstoi's unhappy families, is dystopic in its own way, charting its own unique path to a particular catastrophe. Only the most procrustean criticism will try to lump them together.)

The most successful utopias are those, like *Star Trek*, that are unaware of their didactic burden and so wear it lightly. In the '60s and early '70s, contemporary with *Star Trek*, such unwitting utopias proliferated within SF, particularly among writers associated with the New Wave and the London-based magazine that published their earliest, signature work, *New Worlds*, under the charismatic editorship of Michael Moorcock. I lived in England through much of that period (1965–1970), published a fair portion of my own best SF in the pages of *New Worlds*, got to know all the players, and was accounted one myself.

Literary movements tend to be compounded, in various proportions, of the genius of two or three genuinely original talents, some few other capable or established writers who have been co-opted or gone along for the ride, the apprentice work of epigones and wannabes, and a great deal of hype. My sense of the New Wave, with twenty-five years of hindsight, is that its irreducible nucleus was the dyad of J. G. Ballard and Michael Moorcock, with Ballard in the role of T. S. Eliot, the genius in residence, and Moorcock as Ezra Pound, a Svengali for all seasons, ready to welcome anyone into the club who might in some way advance the cause. They were essential to each other (and to the cause), for without Moorcock and *New Worlds* to beat the drum, Ballard's work would have appeared in only those few avant-garde venues receptive to the transgressive fictions of non-Establishment writers (and far from the madding crowd, or popular success); and without Ballard's conspicuous and then prolific talent to showcase, the New Wave and *New Worlds* would never have reached escape velocity.

What the New Wave offered was science fiction without spaceships. In the Ballardian future, the space program, then in its infancy, was transformed into a gothic ruin, a postapocalyptic Tintern Abbey. In the novels he wrote that predate his New Wave persona, Ballard designed a series of global nonnuclear catastrophes—by wind, by water, by flame—remarkable for their decorative beauty and existential chill. Ballard had a visual imagination, but it was one that had been formed, essentially, by the surrealists Dalí, Ernst, Magritte, and Tanguy—painters who by the mid-'60s were as easy to digest as Norman Rockwell.

The surrealists' strategy, and Ballard's, was the glaring juxtaposition of banal and high-intensity images, a pairing that elevated the banal into the mythological and deflated cultural icons into clichés. Nudes became furniture, and old etchings commingled with ads and clipboard art in a democracy where all images were equal. Here is how Ballard summarized his own New Wave aesthetic in a 1990 footnote to the title story of his 1970 collection, *The Atrocity Exhibition* (which was released by Grove Press in the United States under the lapel-grabbing title, *Love and Napalm: Export USA*):

> "Eniwetok and Luna Park" may seem a strange pairing, the H-Bomb test-site in the Marshall Islands with the Paris fun-fair loved by the surrealists. But the endless newsreel clips of nuclear explosions that we saw on TV in the 1960s (a powerful incitement to the psychotic imagination, sanctioning *everything*) did have a carnival air, a media phenomenon which Stanley Kubrick caught perfectly at the end of Dr. Strangelove.[9]

Bingo! I thought, when I came upon that passage. In those two sentences, Ballard conflates the entire history of postwar SF as it relates to the culture at large, particularly the notion I set forth in Chapter 4 that the Bomb was both unreal, "a media phenomenon" (as in Dick's *The Penultimate Truth*), and a release valve for libidinal forces which themselves are a conflation of high culture (Paris, the surrealists) and low (an amusement park). In a rare salute to one of his contemporaries (Ballard protected his genius by ignoring the work of rival millenarians), Ballard cites the climax of *Dr. Strangelove* (when Slim Pickens rides an H-bomb to rodeo glory) as a locus classicus of the cultural transition from nuclear dread to entertainment, and he even rounds off his footnote with a salute to the higher wisdom of psychosis, that favorite trope of the mid-'60s Age of Aquarius: "*The Atrocity Exhibition*'s original dedication should have been 'To the Insane.' I owe them everything." This, in reference to the exhibition of the title, a show of paintings by the patients of an insane asylum. Poe would have approved.

I met Ballard in spring 1966 on my first stint of expatriation in England. I was twenty-six, with only one U.S. paperback novel to my credit; Ballard was ten years my senior, a widower, and already filled with the afflatus of his own genius. My visits took the invariable form of a trip to the Shepperton train station south of London and then a terrifying ride with Ballard at the wheel of his sports car. At his home, a dilapidated, infinitely cluttered bungalow that he shared with his two children, Ballard, fueled with whisky, would deliver an oral version of his private gospel. Sad to say, I remember not a single oracle from those occasions, only a sense that the man was, as advertised, a genius hard-wired to the Zeitgeist. I did take in the disparity between the slightly raddled Ballard I could see before me and the more glamourous versions of the author who took starring roles in his "condensed novels" under the name of Travis or Travers or Traven, but I'd met enough renowned authors by then not to find that dismaying. He was a brilliant monologuist, and I accepted my role as his listener with as much reverence as if I'd paid to attend a seminar.

The lesson to be drawn from the monologues, and from the fiction Ballard was writing then, was that we were living in the last days of Western civilization and that it was a blast. For Ballard, this was not a theoretical or visionary proposition. He had spent his adolescence interned in a Japanese prisoner-of-war camp in mainland China, where his teenage role models were the apprentice kamikaze pilots at a camp nearby. (The story is told in his 1984 autobiographical novel, *Empire of the Sun*, filmed by Steven Spielberg in 1987.) For Ballard the prospect of violent death was real and thrilling—as it was for anyone in the passenger seat of his sports car. Ballard was acutely, even sadistically, aware of this. "Auto-erotic" was a favorite double entendre, and he once curated a show of wrecked cars for a London art gallery, as a pendant to his novel

Crash (1973). His entry in *The Encyclopedia of Science Fiction* characterizes *Crash* as "perhaps the best example of 'pornographic' SF," a book that "explores the psychological satisfactions of danger, mutilation and death on the roads. . . . One publisher's reader wrote of the manuscript, 'The author of this book is beyond psychiatric help.'"

Earlier I suggested that the automobile is the "secret meaning" of the rocket ship. Ballard, in erasing the rocket ship from his fiction, and along with it the notion of outer space as the new frontier, found a new subject matter for SF: the present in, as it were, its futuristic aspect. He could look at the world around him—suburban Shepperton—with the radical innocence of someone whose home town had been a Japanese internment camp. And everything was strange. The sports car that he owned and drove like a kamikaze pilot was a good deal stranger and more vivid than any rocket ship, which existed, if at all, only as a TV image among a host of other images. Which of those images was to be privileged as "reality"? Why not, simply, those that *he* found most exciting, the spaceships in his own backyard? Why not build a future from those images rather than from the do-it-yourself kits of traditional SF? In answering those questions to his own personal satisfaction, Ballard invented Inner Space.

Like the inventions of so many mad scientists in the annals of sci fi, it was to have unforeseen consequences once it escaped from the inventor's laboratory. Quite simply, Inner Space became shorthand for sex, drugs, and rock 'n' roll, that troika of '60s hedonism for the masses.

At one level, for the more upscale Ballard aficionados, the New Wave represented science fiction's coming of age, as it appropriated existing literary techniques of the avant-garde as practiced by those writers one had studied in college. A new breed of SF readers and writers were hungry for the literary equivalent of Julia Child. I am probably typical of that new breed. Born in 1940 into a lower-middle-class family, I was able to attend college at a time when universities had become a growth industry. The writers I was taught to respect and emulate in those years were the early modernists: Joyce, Kafka, Mann, Camus, plus such new arrivals as Beckett, Genet, and Pinter. If one came to such authors having grown up in the ghetto neighborhood of SF, how could one not aspire, if one returned to the hood, to bring one's education along? Dozens of us tried to do just that: myself, Norman Spinrad, Joanna Russ, Harlan Ellison, John Sladek,

Ursula Le Guin, Gene Wolfe. Virtually the entire U.S. component of the New Wave shared a common missionary goal of wanting to bring some literary couth to a genre we all regarded as exciting but basically déclassé, with honorable exceptions like Clarke, Dick, and Bester.

Science fiction, however, isn't about literature. It is a form of popular entertainment, part of what was then being assimilated into the national "entertainment industry," an industry that has since become international in scope. And the reason that the New Wave made the mark that it did was that Michael Moorcock, the editor of *New Worlds* and the P. T. Barnum of the New Wave, understood that essential fact. Moorcock, unlike all the other writers associated with the New Wave, had not gone to college. Instead, at age eighteen, he became the editor of *Tarzan Adventures* and began to produce, in Balzacian abundance, pulp fiction of ever increasing proficiency. He first made his mark as a commercial fictioneer with a series of sword-and-sorcery fantasies featuring Elric of Melnibone, a hero possessing, and possessed by, a phallic sword with a will of its own. As Moorcock matured, Elric metamorphosed into Jerry Cornelius—and the editor of *Tarzan Adventures* became the editor of *New Worlds*.

Moorcock understood that the most important attribute of any literary protagonist who hopes to have mass audience appeal is the right wardrobe, one that his fans will want to wear themselves, if only at the Mardi Gras ball of their daydreams. *Star Trek* fans had their pajamas. This is how Jerry Cornelius appeared in his 1965 debut in *New Worlds*:

> In his permanently booked room in The Yachtsman, Jerry Cornelius had awakened at seven o'clock that morning and dressed himself in a lemon shirt with small ebony cufflinks, a wide black cravat, dark green waistcoat and matching hipster pants, black socks and black handmade boots. He had washed his fine hair, and now he brushed it carefully until it shone.
>
> Then he brushed one of his double-breasted black car coats and put it on.
>
> He pulled on black calf gloves and was ready to face the world as soon as he put on his dark glasses.
>
> From the bed, he picked up what appeared to be a dark leather toilet case. He snapped it open to check that his needle gun was pressured. He put the gun back and closed the case.

> Holding the case in his left hand, he went downstairs; nodded to the proprietor who nodded back; and got into his newly polished Cadillac.[10]

This manual on a perfectly modern lifestyle continues with an explanation of how Jerry gets along on a diet of Mars bars, coffee, pills, and Bell's scotch. Jerry is a proletarian James Bond, outfitted by Carnaby Street (where one might then buy all the clothes he modeled), and ready for every transgression. He is as old as Don Juan and as young as Gangsta Rap. He is Everyman stripped to his Id, and he is meant to be winkingly understood as such, by even the most naive reader. Cornelius was an inspired invention and a commercial and critical success. *The Final Programme*, the first Jerry Cornelius novel, was filmed in 1973, and the fourth, *The Condition of Muzak*, won the *Guardian* Fiction Prize in 1977, an unprecedented achievement for a work of genre fiction.

Moorcock was one of the most prolific authors in a genre noted for high productivity. In addition to the four Jerry Cornelius novels, there were story collections, spinoffs that featured ancillary characters, and an anthology of Jerry Cornelius stories by other *New Worlds* writers. Now, three decades later, the books are mostly out of print, and the fashions of Carnaby Street are flea market rags. But the truth of Jerry Cornelius goes marching on.

SF is in its nature an ephemeral literature. Most predictions of the future are wide of the mark, and their errors become more glaring as the years progress. No blossoms wither so quickly as yesterday's tomorrows. Even so, Moorcock was utterly prescient in his sense of what the Zeitgeist demanded of SF in the Age of Aquarius: not the panoramas of megaengineering that reached their apotheosis in the space program and its fictional offspring like *2001*; rather, a celebration of the audience's daydreams, of Inner Space and future fashions—a lifestyle and a mind-set any lad might hope to buy once he'd sat his way through school and found a job.

Moorcock's dedication of *The Final Programme* reads like a manifesto for this new kind of SF: "To Jimmy Ballard, Bill Burroughs, and the Beatles, who are pointing the way through." Ballard is there, necessarily, for collegial and strategic reasons, while the Beatles, then at the zenith of their success, represent the author's own dearest dream, that he might aspire to a cognate success as the age's premier "Paperback Writer," a song that became, the moment it was released, the national anthem of all SF writers. But William Burroughs? How does he come to figure among this new trinity of B's? He was not a *New Worlds* contributor nor, yet (though the dedicator calls him "Bill"), an intimate friend; he was not even one of the authors whose style was imitated by other New Wave writers. Why then?

Because his name was a synonym for drugs, and drugs—illicit but universally available—were the distinctive transgressive feature of the future foreseen by New Wave SF. A few paragraphs earlier, I spoke of the troika of sex, drugs, and rock 'n' roll as being central to the '60s Zeitgeist, but it was drugs, alas, that were to be most central. Sex, after all, must be part of *any* Zeitgeist. Baptists, Republicans, Zoroastrians, Flat-Earthers, Afrocentrists, members of Mensa and the ACLU: they all have sex, and think about it when they're not having it, and (once the Supreme Court had given the go-ahead) write about it when they write novels. Sex is universal. And rock 'n' roll? As the song says, it's here to stay. But if it should ever fade away, there will always be some kind of music to replace it—polkas or salsa or ska. Something energetic, anyhow, to dance to and sing along with. Music is universal too.

But drugs? Drugs, in any sense that Burroughs can figure as a metonym, are another matter. Few writers before the '60s were upfront about their own use of the scarier drugs (De Quincey and Cocteau are notable exceptions), from a prudential wish not to be thought *poètes maudits* like Poe or Rimbaud. However, in the '60s, for a certain kind of writer, disgrace became a career possibility, and in the arena of can-you-top-this? confessionalism, no one could hold a candle to Burroughs, who was not only embalmed in heroin through most of his adult life but had murdered his wife—and, like O. J. Simpson, got away with it. He was openly queer long before most other gays uncloseted themselves, and his published erotic fantasies dwelled obsessively on pedophilic rape and murder. As a transgressor, Burroughs was in a class with Gilles de Retz.[11]

If Burroughs's advocacy of drug use had been typical, readers would have been as little likely to experiment with drugs as with white slavery or thrill killing, pleasures that most of us are content to enjoy vicariously in pulp fiction or at the movies.[12] The true Pied Piper of the Drug Age was Aldous Huxley, whose 1954 memoir of his own experiences with mescaline, *The Doors of Perception*, was to become required reading for the '60s counterculture. In *Brave New World* (1932) Huxley had already celebrated an imaginary drug, soma, a cocaine-like drug responsible in that novel for much mindless euphoria and one lethal overdose. Soma, being unreal and so unobtainable, can't lead Huxley's readers into temptation, but the author's account of his mescaline experiences is advertising copy of the highest order. Who could resist the opportunity to enter "the doors of perception" if just across the threshold is a flower arrangement like this? Three blossoms, a rose, a carnation, and an iris are transformed into a transcendent vision of

> nothing more, and nothing less, than what they actually were—a transience that was yet eternal life, a perpetual perishing that was at the same time pure Being, a bundle of minute particulars in which, by some unspeakable and yet self-evident paradox, was to be seen in the divine source of all existence. . . . The Beatific Vision, *Sat Chit Ananda*, Being-Awareness-Bliss—for the first time I understood, not on the verbal level, not by inchoate hints at a distance, but precisely and completely what those prodigious syllables referred to.[13]

Hey—thought I, reading those lines in the early '60s; let me have some of that *Sat Chit Ananda*. And in the spring of 1966, I was able to act on that thought when I was given the opportunity to enter the doors of perception myself—not via mescaline but with a tab of its synthetic equivalent, LSD. Huxley had given permission, and though I cannot be sure that the experience was precisely and completely what the prodigious syllables referred to, it certainly was Lucy in the Sky with Diamonds. At the height of that first trip, having hiked to the ruins of a medieval fortress near Fuengirola, in Spain, I could survey a wide prospect of the Mediterranean, across the surface of which, in living Arabic script of wave foam, I read the ineffable—if ultimately indecipherable—Text of the World. A hallucination, assuredly, but so beautiful! And even more beautiful, upon the mottled walls of the fortress, open to the sky, I could view, as upon a movie screen, an animated tapestry of rare device: an Oriental marketplace bustling with activity, as though an Indian miniature had come to life, with the neon-intense pastels peculiar to that art form. I marveled, not so much at this view of Xanadu, but at the conjoint powers of memory and imagination that could project, from my mind, to that soot-stained, ruined wall-cum-movie-screen, such a vivid and detailed panorama. I knew, as I watched it, that what I saw was, in fact, what I might see any night in an especially good dream; that somehow the drug had dissolved the membrane that separates perception and fancy, allowing me to be a spectator of my own pleasure principle on holiday.

Does that amount to a confession? Does Spain have a statute of limitations going back to 1966? Then I'm a guilty wretch—not only then and there, but for later acid trips in the States and in England, the last of which took place in rural Surrey in 1971, where a field of barley became as menacing as one of Van Gogh's last, minatory paintings, and I knew that it would behoove me to close the doors of perception until such time as I knew they would not open, inadvertently, into another such subjective hell.

For a good many of my generation, that is a common tale. I relate it neither to claim accreditation as a graduate of the Zeitgeist nor to attest to some penitential hindsight, but because I don't think it would be proper to speak of the connection between SF and the '60s drug culture as though I had not been there, sharing the pleasure and the risks.

The risks were high, as has become increasingly apparent, although SF writers have had a lower mortality rate than rock stars, for perhaps the simple reason that paperback writers earn lower salaries and could not binge on pricey drugs to the same degree. Michael Moorcock, for instance, was no Jim Morrison; his wild excesses existed mostly on paper. Ballard drank, and drove, but drugs? I doubt it. In America, the writers of the New Wave were even more temperate. Harlan Ellison was a virtual Mormon in that respect; liquor never touched his lips, never mind other controlled substances. Even Phil Dick, SF's major authority on the subject of drug abuse, died, ironically, from the *lack* of a drug: the blood pressure medication that he did not use with prescribed regularity.

Marijuana was another matter. Grass was everywhere. On my first science-fictional social occasion, at a party at the home of Terry Carr, my then agent and later a sometime editor, grass and the Beach Boys provided the ambience. Grass was there at the first meeting of Paul Williams and Phil Dick (and at my first meeting with Dick, years later, as well). It was there at every SF convention I ever attended, as the alternative, grown-up indulgence for those who didn't dress up in *Star Trek* pajamas for the masquerade contests. It was there the one time I met Theodore Sturgeon and was persuaded to sample the nudist lifestyle and invited to stay overnight for a threesome with Mrs. Sturgeon (an opportunity I declined). It was there in New York, London, L.A., and in every provincial Holiday Inn that ever hosted a science-fiction convention. And it was, a lot of the time, a lot of fun. As Dylan sang, everybody must get stoned.

But it did gradually become apparent that confirmed potheads tended to lose their edge. Dick was a notable exception, and many of those who constitute the rule are still alive, and so to catalogue them would be a needless cruelty. But if you're curious, read between the lines of those senior writers who once seemed so wonderful and who now, so noticeably, are not. The reason, when it isn't booze, is probably pot.

SF mirrors the rest of contemporary literary and pop culture in its ambivalent relationship to drugs. Those novels targeted for the cyberpunk audience are almost required to celebrate chemical holiday making, though they often will present the theme in disguise—as adventures in nanotechnology (microscopic "machines" that mess with your mind) or virtual reality (software that messes with your mind). Since the '60s SF has prided itself on being, in itself, so "trippy" as to obviate the necessity for other stimulants, but that is just wishful thinking or PR. So long as drugs are commonly available and widely used, the temporary transcendance they offer will always appeal to those looking to have their minds blown. And you can't Just Say No to that element of human nature that would prefer to say Far Out! without going after the imagination root and branch.

Chapter 6

CAN GIRLS PLAY TOO? FEMINIZING SF

Sci fi of the pulp era (1929–1956) was largely a male enclave. However, as contemporary feminists are wont to point out, *all* literature from the time of Homer to that of Sylvia Plath has been a male enclave. Sci fi may have been somewhat more egregiously segregated in those days, because it was so déclassé, only a step above comic books, that few women would have been tempted to fight for rooms of their own in such a tawdry residence. The first women who did crash the party—notably C. L. Moore and Judith Merril—often did so (like Mrs. Browning or Sylvia Plath) as spouses. Often, like both Moore and Merril, they published under male psuedonyms.

Women were also neglected, libeled, or condescended to when they appeared *in* science-fiction stories. Typically they were damsels in distress, like Buck Rogers's lady friends of the twenty-fifth century, Willma Deering and Princess Alura of Mars. (Merril, incidentally, was one of that comic strip's many writers.) A sign of just how achingly gauche the SF of that era could be is the "classic" status enjoyed by Lester del Rey's "Helen O'Loy," the story (see the Introduction) that first posited the equation, Housewife = Robot.

Apologists for SF of the pulp period often contend that the genre's practice of simply ignoring women has been a practical strength that helped separate SF from mundane literature, which dealt primarily with gossipy trivia like courtship and suburban adultery, while SF celebrated the Destiny of Man in the new frontier of outer space, where women were shown to be incompetent ditzes. Another classic tale of 1954, "The Cold Equations" by Tom Godwin, illustrates what is likely to happen to girls when they trespass into the boys' club of Outer Space. An astronaut delivering a vital supply of serum to cure an epidemic on a far planet discovers that a teenage girl is a stowaway aboard his ship. The extra kilograms of her presence means the spaceship will not have fuel enough to land, and so, no matter how much she whines (for she can't grasp the "cold equations" of the title), she must be jettisoned.

One need not be a feminist to feel that stories like "Helen O'Loy" and "The Cold Equations" might give offense to female readers. But that was not a source of controversy, because women, by and large, were as little tempted to read pulp SF as they were to read men's adventure magazines or hard-boiled detective stories. Readers of that era lived in two different worlds, clearly labeled Gents and Ladies, and they were as little likely to trespass into each other's reading domains as they were to enter bathrooms similarly coded.

Then, in 1959, the world of mundane literature underwent a revolution whose impact would soon be felt in all adjacent territories, from SF to soap opera. The Supreme Court decided that the post office had not been within its rights in banning *Lady Chatterley's Lover* as pornographic, and in very little time all the classic stiffeners were in print and on sale across the country, from *Fanny Hill* to *Tropic of Cancer*, plus envelope-pushing new work by literary luminaries (Philip Roth, Erica Jong) and not-so luminaries. Among the latter were a few seasoned SF hacks, who, in the boom years of the '60s, produced reams of legal porn on automatic pilot. The real speed demons, like Robert Silverberg and Barry Malzberg, could produce entire books in a single weekend.

The new freedom was soon claimed by writers of all genres and every level of brow. The vanguard of erotic realism had been works that could appear before the Supreme Court belaureled with blurbs from eminent literary professionals. Judge Bryan's decision in the *Lady Chatterley* case made much of such bona fides:

> The Grove edition [of *Lady Chatterley's Lover*] has a preface by Archibald MacLeish, former Librarian of Congress, Pulitzer Prize winner, and one of this country's most distinguished poets and literary figures, giving his appraisal of the novel. There follows an introduction by Mark Schorer, Professor of English Literature at the University of California, a leading scholar of D. H. Lawrence and his work. . . .
>
> There is nothing of "the leer of the sensualist" in the promotion or methods of distribution of this book. There is no suggestion of any attempt to pander to the lewd and lascivious minded for profit.[1]

This technique of innocence by association became standard operating procedure. Robert Lowell pinned this flower on the jacket of Burroughs's *Naked Lunch:* "It's a completely powerful and serious book. . . . I don't see how it could be considered immoral." A decade later, Burroughs himself was testifying as to the merits of J. G. Ballard's *Love and Napalm: Export USA*, as his story collection, *The Atrocity Exhibition,* had been retitled for U.S. consumption: "A profound and disquieting book. The nonsexual roots of sexuality are explored with a surgeon's precision. An auto crash can be more sexually stimulating than a pornographic picture. . . . This book stirs sexual depths untouched by the hardest-core illustrated porn." Talk about praise from Caesar.

Ballard cannot be held responsible for the way his publisher chose to promote his work, but Judge Bryan's benign assumption that no one is hoping to profit by an appeal to the lewd and lascivious minded begins to sound a bit disingenuous. The plain fact is that many writers, from Flaubert to Mickey Spillane, have exerted all their writerly resources to appeal to the lewd and lascivious minded, and for at least two compelling reasons: (1) they are lewd and lascivious minded themselves and find some gratification in sharing their erotic fantasies with those of a similar disposition, and (2) they stand to make good money by doing so.

It was not long before the full-frontal sexual candor of Ballard and others of a New Wave ilk had been adapted to the needs of the larger, fannish audience of nerdy teenagers and those older readers whose taste in reading was dictated by the nerd within. Enter John Frederick Lange, Jr., who under the pseudonym of John Norman, wrote a series of planetary romances, beginning with *Tarnsman of Gor* in 1966 and continuing until 1988, with, among others, *Assassin, Captive, Slave Girl, Fighting Slave, Savages, Mercenaries, Dancer,* and *Renegades of Gor.* As those titles should telegraph, and the cover art made even clearer, Lange was a bondage freak. In book after book, he retailed the same masturbatory fantasy. This is how it goes:

> "On Gor," whispered the girl next to me, "I have learned that men and women are not identical. . . . And men, or Gorean men, or men of a Gorean type are the masters."
>
> "Yes," I said.
>
> "And women such as I are their slaves," she said.
>
> "Yes," I said. "Lick and kiss me." . . .
>
> "You command me like a Gorean slave girl," she said.
>
> "That is what you are," I told her.
>
> "Yes, Master," she said.
>
> [After the happy slave has been privileged to please her master, their dialogue continues:]
>
> "You treated me like a Gorean slave girl," she said.
>
> "That is what you are," I told her.
>
> "Yes, Master," she laughed. "It is true. . . . I am the same. I am no different. I am only another girl in the collar, another woman who must obey you and serve your pleasure."
>
> "Are you content?" I asked.
>
> "Yes, Master," she said, "as would be any woman in the arms of a man such as you. . . . At last I have come to a world where there are men who wish for me to please them, and will see that I do so, and want me, and will have me, a world where there are masters."
>
> "I must be going," I told her.[2]

Before sneering, consider whether, if the same sentiments were expressed in a more nuanced prose style—by the author, for instance, of *The Story of O*—you would react similarly. The Gor books are addressed to a Budweiser audience, O's story to those who prefer Côtes du Rhone; such literary attention as Lange may receive is limited to derision or reprobation, while the creator of O became an idol of the French literati. When the upper classes practice kinky sex, it is merchandised as luxury goods, like Mapplethorpe photographs; when the lower classes do the same thing, censors begin to worry for the unraveling fabric of civilization.

Science fiction, in its pulp magazine heyday, was a playground for younger readers and for those willing and able to regress to a state of blinkered innocence. Before they realized that the new candor might accommodate their own writerly impulses, SF writers tended to reprobate sexual candor. Robert Heinlein, for instance, had maintained that one reason SF was superior to mainstream literature was that it did not pander to low desires and lascivious imaginings. But within a year of the *Lady Chatterley* decision, Heinlein was at work on *Stranger in a Strange Land*, which even in the slightly expurgated version of 1961 celebrates nudity and free love in a spirit for which Judge Bryan's phrase, "the leer of the sensualist," is entirely apt.

Heinlein's hero in *Stranger* is Valentine Michael Smith, an average American boy and a foundling, who, having been raised by Martians, has Superman-like psychic powers. He can kill his enemies by wishing them out of existence. Girls find him irresistible, and, unlike the Clark Kent/Superman of that era, he found girls irresistible. This gives rise to hours of earnest discussion on what is to be done in such a situation, just as it has in freshman dorms time out of mind. There are even glimpses of what, until that time, SF had scarcely dared mention. The most explicit such scene had to be cut from Heinlein's first draft at his publisher's insistence, but it was subsequently restored in the Ace/Putnam "original uncut version" of 1991. It takes place after Mike (our hero) and one of his consorts, Jill, have been explaining, in veiled terms, the benefits of their commune-cum-love-nest to a potential recruit, Ben:[3]

> "Ben, you won't believe it until you've had it done for you, but Mike can lend you strength—physical strength, I mean, not just moral support. I can do it a little bit. Mike can really do it."
>
> "Jill can do it quite a lot." Mike caressed her. "Little Brother is a tower of strength to everybody. Last night she certainly was." He smiled down at her, then sang:

You'll never find a girl like Jill.
No, not one in a billion.
Of all the tarts who ever will
The willingest is our Gillian!

"—isn't that right, Little Brother?"

"Pooh," answered Jill, obviously pleased, covering his hand with her own and pressing it to her. "Dawn is exactly like me and you know it—and every bit as willing."

What all this tenderness portends is finally revealed in the passage cut from the 1961 edition. As Ben explains to his mentor, Jubal:

> "Jubal," Caxton said earnestly, "I wouldn't tell you this part at all . . . if it weren't essential to explaining how I feel about the whole thing. . . . By this morning I was myself half conned into thinking everything was all right—weird as hell in spots, but jolly. Mike himself had me fascinated, too—his new personality is pretty powerful. Cocky and too much supersalesman . . . but very compelling. Then he—or both of them—got me rather embarrassed, so I took that chance to get up from the couch.
>
> "Then I glanced back—and couldn't believe my eyes. I hadn't been turned away five seconds . . . and Mike had managed to get rid of every stitch of clothes . . . and so help me, they were going at it, with myself and three or four others in the room at the time—just as boldly as monkeys in a zoo!
>
> "Jubal, I was so shocked, I almost lost my breakfast."[4]

Ben here is serving as a spokesman for those readers for whom Michael's monkey business cannot be contemplated without reprobation. In this, and in his use of polite euphemisms when dealing with sex, Heinlein is still a relative bulwark of propriety, but his message is not essentially that different from Lange's in his Gor books. Jill is "obviously pleased" to be memorialized in song as the willingest of all the tarts. *Stranger in a Strange Land*, together with later Heinlein novels combining soft-core porn with SF, also resembles the work of Lange in devoting long entire chapters to theories of sex that are the male chauvinist equivalent of Andrea Dworkin's conviction that all men are rapists and all intercourse between the sexes criminal in intent. In the Lange/Heinlein formulation, it is a woman's genetic destiny to serve man's pleasure and to submit to his will. The interminable bull sessions in their novels in which the theory of male supremacy is set forth serve two purposes: (1) they pad out the word count and (2) reassure the presumably reluctant reader that it is all right to enjoy such fantasies, in much the same way that a professional sex worker offers her clients reassurances as to their puissance and her pleasure. That Heinlein understands the necessity for such assistance to male self-esteem is clear from a passage in his 1982 novel, *Friday,* in which the title character, a professional killer for the U.S. government, explains how she survived gang rape by pretending to enjoy it:

> I worked on all of them—method acting, of course—reluctant, have to be forced, then gradually your passion overcomes you; you just can't help yourself. Any man will believe that routine, they are suckers for it.[5]

At the conclusion of the novel, as a treat for those readers who might have felt betrayed by Friday's playacting, Friday falls in love with her principal rapist.

Friday, a rather routine futuristic thriller except that the starring, James Bond–like role is given to a woman, may be considered Heinlein's mischievous response to feminism, which by 1982 had made a dent in the consciousness of even the most cloistered SF readers. The book's dedication page offers a tip of the Heinlein hat to Anne McCaffrey, Vonda McIntyre, Ursula Le Guin, and twenty-eight other women (identified by first names only), all of them SF professionals or the spouses thereof—a gesture as cozy and redolent of the mind-set of SF writers of Heinlein's generation as Maurice Chevalier's rendition of "Thank Heaven for Little Girls."

By 1982 the field of science fiction had undergone an enormous expansion and diversification from its condition of twenty years earlier. A few elder statemen, like Arthur Clarke and Frederik Pohl, were still producing books that were not too transparently the literary equivalent of a repetition compulsion, such as had come to afflict Bradbury, Heinlein, Asimov, and Herbert in the years of their affluent decline. For the most part, however, the field had become a loosely knit league of independent fiefdoms, and one of the most considerable of these was (to borrow a phrase from one of its future satraps, Sheri Tepper) Women's Country.

The pioneers of Women's Country—women writers of the '70s, preeminently Joanna Russ—tended to postulate planets where heroines with black belts in karate could rumble with assorted forms of male chauvinist pigs, then take a quick tumble in the hay, and proceed on their way to fresh adventures in the spirit of picaresque daydreaming that boys of all ages had been enjoying since the days of chivalric romance some centuries earlier. That's how Heinlein's hit-person Friday got her rocks off, and it is also, with some modifications, the formula for space operas by the three Heinlein dedicatees mentioned above. Feminism had made enough headway by the '70s that even fogeys could see the justice of equal opportunity daydreaming. If guys can have their Beowulfs and Rolands and Captain Videos, why shouldn't girls enjoy their own warrior imaginings—especially if that could mean doubling the size of the SF audience. Justice and the bottom line combined to produce an unstoppable market force, and now even the latest *Star Trek* spaceship is commanded by a woman—as are, increasingly, units of the armed forces.

The three writers who have been most successful in recruiting women readers for SF—Anne McCaffrey, Vonda McIntyre, and Ursula Le Guin—have done so by virtue of their skill in retooling SF conventions to the needs of an audience alert to feminist issues but also hungry for escapist fantasies tailored specifically to the female imagination. Anne McCaffrey, who has been the most successful, and least politically correct, of the three, resembles one of her own can-do heroines. She is a former opera singer of robust build and great vitality, an avid horsewoman, and an indefatigable novelist whose books have earned her a stately home in Ireland. Her most popular and extensive series concerns the lost Earth colony of Pern, where young women may adventure about on time-traveling, telepathic, fire-snorting dragons. Pern has become the favorite fantasy destination of younger female readers who once would have had to be content with mundane novels about girls on horses. McCaffrey's novels offer the girls who read them vicarious satisfactions equivalent to the SF that caters to boys, but this equal opportunity has a glass ceiling. Just as in the Nancy Drew books, the basic conventions and protocols of gender (pink for girls, blue for boys) are maintained. McCaffrey's women know their place.

I remember attending a workshop at the Milford SF Writers Convention in the late '60s at which Anne had asked her fellow authors' advice about an as yet unsold story revolving around the decision of a young woman to put aside her ambitions as a wielder of psychic powers in order to follow the higher calling of motherhood. This story line was sniffed at by some of the assembled writers as a thinly disguised appeal for women to exit the job market and return to their traditional roles as wives and mothers. And why not, if that's the story McCaffrey wanted to tell? The problem with the story wasn't the message as such but the fact that it was not delivered with any imaginative or emotional oomph. One of the writers present suggested that Anne's heroine could have her cake and eat it too by the simple expedient of discovering herself to possess a new and higher psychic talent: she could telekinetically manipulate DNA at the moment of conception so as to enhance her future offspring's psychic powers. The rewritten story was soon thereafter featured on the cover of *Analog*.

Vonda McIntyre's target audience is older than that of Anne McCaffrey, but it too dwells primarily within the borders of Women's Country. The McIntyre Weltanschauung divides everyone into an uncaring, imperceptive, close-minded Them (the patriarchy) and a loving, hip, holistic, and victimized Us (guess who?). Pain, openly acknowledged, is a badge of pride. In one story after another, McIntyre's characters are implored to surrender to their stifled need to cry, by way of showing themselves to be one of Us. An instance, from her Nebula Award–winning story, "Of Mist, and Grass, and Sand":

> "Can any of you cry?" she said. "Can any of you cry for me and my despair, or for them and their guilt, or for small things and their pain?" She felt tears slip down her cheeks.
>
> They did not understand her; they were offended by her crying.[6]

I like to think that I can wet a handkerchief with the best of them, but I do find such earnest solicitations to tears not only tasteless but bullying, as though Orwell's Thought Police had been taking sensitivity training.

In fairness, it must be said that when she is not conducting a group therapy session, McIntyre can write with a spare, modulated prose that bears comparison with that of her mentor, Ursula Le Guin. The awards on her trophy shelf are well merited. John Clute has put the case for McIntyre and the other denizens of Women's Country with characteristic critical generosity:

> SF writers are now free to assume competent female protagonists, and in this year [1978] Charnas, McIntyre, Russ, Tiptree, and Yarbro all do so. Vonda McIntyre's *Dreamsnake* [a novel that evolved from the story quoted above] is perhaps the most interesting: female protagonists who are empowered by their capacity to heal are on delicate ground—it is far too easy (especially for men) to sentimentalize such healing women into walking comfort stations. But the icon of healer is potent, and in a literary sense it has a healing effect on the SF genre as a whole. Partly because of books like *Dreamsnake*, it is no longer possible for SF authors to pretend that a rounded society can be properly described simply in terms of the wars that it wages and the frontiers that it penetrates.[7]

The awards McIntyre won for her early fiction (a Nebula for "Of Mist, Grass, and Sand," another Nebula and a Hugo for *Dreamsnake*) positioned the author for a novel form of success.[8] She was chosen to write the novelizations of three *Star Trek* movies and two other books in the same franchise. That she could do so without any aesthetic or ideological compunctions is a reflection of the shifting cultural positioning of SF in the 1980s, as genre expectations converged on something a bit higher than the lowest common denominator of earlier televised SF, and as *Star Trek*, along with other TV series, adjusted its scenarios to a moderate form of political correctitude.

The most successful, and the most significant, feminist presence in the SF field has undoubtedly been that of Ursula Le Guin. Although her books may not initially have sold in mass quantities like those of

McCaffrey and McIntyre (when in *Star Trek* mode), two of them, *The Left Hand of Darkness* (1969) and *The Dispossessed* (1974), have attained classic status. She's won five Hugos and four Nebulas, and she commands an unrivaled respect in those academic circles that pay attention to the genre. She is the most respectable SF writer going.

Respectability has its price. One does not read Le Guin for fun, or excitement, or wild ideas, nor yet for what is often accounted SF's raison d'être, a sense of wonder. Her most renowned novel, *The Left Hand of Darkness*, is equal parts an anthropologist's field report about a planet, Winter, inhabited by humanoid hermaphrodites and a diary of a journey across a glacier shared by an indigene, Estraven, and the male human narrator, Genly Ai. In a world of violence, court intrigues, and untrammeled sex, Estraven and Genly Ai maintain a chaste relationship and practice a Gandhian passive resistance. Facing defeat and ignominy, Estraven chooses a heroic death in the form of skiing downhill into enfilading machine-gun fire. At the moment of their greatest intimacy, Genly Ai teaches Estraven to communicate by telepathy, but this moment of shared intimacy is without further repercussions on the plot. Rather, it signposts the book's feminist good intentions. In the '60s and '70s men fought and flew rocket ships; women were telepathic and performed psychic healings—whether one were aboard the Starship *Enterprise* or reading the novels of McCaffrey, McIntrye, or Le Guin. Le Guin especially offered the slenderest rations of vicarious excitement. She is an educator, her aim to instruct and edify, and the focus of her instruction, over a writing career that now spans three decades, has narrowed down to a single issue: feminism.

Le Guin's feminism is less overtly phobic of the male sex than that of Andrea Dworkin, but it is no less absolute. She requires nothing less, if one credits her utopian romance of 1985, *Always Coming Home*, than the abolition of Western civilization as we've known it and the (re)institution of a benevolent, holistic, shamanistic matriarchy. Science fiction, with its opportunity to posit other worlds designed to showcase one's own ideological convictions, is a godsend to any polemicist. Lange had his Gor, and turnabout is fair play. Le Guin, in her Hugo and Nebula Award–winning novella of 1972, *The Word for World Is Forest*, has her Athshe, an edenic planet that has been usurped by barbarous Earthmen who would seem to have received their basic training on Gor. Here is how Le Guin portrays her viciously patriarchal protagonist, Captain Davidson, as he daydreams of laying waste the virgin planet of Athshe:

> Just take up a hopper over one of the deforested areas and catch a mess of creechies [creatures/indigenes/aliens] there, with their damned bows and arrows, and start dropping firejelly cans and watch them run around and burn. It would be all right. It made his belly churn a little to imagine it, just like when he thought about making a woman, or whenever he remembered about when that Sam creechie had attacked him and he had smashed in his whole face with four blows one right after the other. . . .
>
> The fact is, the only time a man is really and entirely a man is when he's just had a woman or just killed another man.[9]

At the time she wrote *The Word for World Is Forest,* Le Guin's polemical target was not so much man as the U.S. presence in Vietnam. Her "creechies" were the Vietnamese peasants being napalmed in their villages, and Captain Davidson was the embodiment of U.S. bloodguilt. But the war came to an end, while the book stayed in print, and Captain Davidson would come to have a more general significance, no longer figuring forth a particular historic wrong but something innate in the nature of men, a sex-specific mark of Cain.

Le Guin's feminism comes bundled with an entire political and economic agenda for the reform of the world in general and science fiction in particular. At an SF convention in Bellingham, Washington, in 1973, Le Guin delivered the first of several pronouncements on the subject, in which she declared:

> The women's movement has made most of us conscious of the fact that SF has either totally ignored women, or presented them as squeaking dolls subject to instant rape by monsters . . . or, at best, loyal little wives or mistresses of accomplished heroes. Male elitism has run rampant in SF. But is it only male elitism? Isn't the "subjection of women" in SF merely a symptom of a whole which is authoritarian, power-worshiping, and intensely parochial?

> ... The only social change presented by most SF has been toward authoritarianism, the domination of ignorant masses by a powerful elite. ... Socialism is never considered as an alternative, and democracy is quite forgotten. Military virtues are taken as ethical ones. Wealth is assumed to be a righteous goal and a personal virtue. Competitive free-market capitalism is the economic destiny of the entire Galaxy. In general, American SF has assumed a permanent hierarchy of superiors and inferiors, with rich, ambitious, aggressive males at the top, then a great gap, and then at the bottom the poor, the uneducated, the faceless masses, and all the women. The whole picture is, if I may say so, curiously "un-American." It is a perfect baboon patriarchy, with the Alpha male on top, being groomed respectfully, from time to time, by his inferiors.[10]

This is tarring with a very broad brush. While there are certainly SF writers whose heroes might have served Le Guin as an inspiration for her Captain Davidson (notably, Heinlein and his posse of imitators), a majority of the most widely read writers would have trouble slipping their feet into Davidson's jackboots. Ever since Edward Bellamy wrote out the specs for the welfare state he longed to see built, SF has been a magnet for writers of left-leaning tendencies. As to the disproportionate attention paid to the rich and powerful, SF is scarcely unique in that regard. Le Guin's main role model, Virginia Woolf, might be scolded for the same failing, if such it is. Even fairy tales and myths might be reprobated by a strict application of Le Guin's principles: Cinderella would be a gold digger and Homer the poet laureate of alpha male baboons.

Ideology breeds nonsense and, in the second and third generation, pernicious nonsense. In the course of the two decades since the publication of her most accomplished novel, *The Dispossessed* (1974), while Le Guin's work has undergone a gradual PC ossification, her reputation among feminist academics has increased by inverse proportion. Marleen S. Barr, of the Virginia Polytechnic University, Blacksburg, has published an essay celebrating Le Guin's collection of linked tales, *Searoad: The Chronicles of Klatsand* (1991), that must have made even Barr's idol cringe with embarrassment. The title of Barr's essay is "*Searoad Chronicles of Klatsand* [sic] as a Pathway toward New Directions in Feminist

Science Fiction: Or, Who's Afraid of Connecting Ursula Le Guin to Virginia Woolf?"[11]

Barr's essay is a marvel of Orwellian duckspeak, the unthinking repetition of the same set phrases with monotonous, unmodified regularity. Her favorites are there in the title: "new directions" and "feminist science fiction," or, as variants, "feminist fabulation" and "feminist thought experiments." In only eight pages, that formula is repeated thirty-three times, as in the following examples: "*Searoad* explores new directions for feminist science fiction which point the way toward ending the sharp distinction between denigrated feminist science fiction and exalted mainstream literature." And: "Through Rosemarie, Le Guin addresses the point that feminist science fiction needs new directions—that it has been given the wrong name." And: "The new direction for feminist science fiction involves moving toward feminist fabulation as the common ground between the groups which Rosemarie, the librarian, and Antal represent." And: "The best new direction for feminist science fiction is for this literature to proceed in a way which nullifies marginalizing the feminist thought experiments which articulate sea changes for patriarchal constructions." And to what conclusion does all this jargon tend? To this: "Although the sun of respectability never sets on the mainstream literary empire, feminist science fiction can strike back. Feminist critics can rest their penchant for overlooking the fantastic. They can see fit to agree with my opinion that Le Guin is the Virginia Woolf of our day."

No author should be held accountable for the zeal of his or her most besotted admirers. Even so, Le Guin's claims for equal time, and laurels, with Virginia Woolf are not the invention of Marleen S. Barr. They are implicit in the story line of *Searoad*, in which a transparently self-reflecting character, Virginia Herne, embodies the author's own sense of her hereditary claim to Woolf's literary laurels, a claim that may be traced back to her essay of 1976, "Science Fiction and Mrs. Brown," in which she takes on the role as Woolf's emissary to the little world of SF.

The sense of grievance expressed by Le Guin, and Barr on her behalf, are characteristic of feminism in its more territorial moments. Feminist demands for equal consideration play a crucial role not only in academia, where they determine appointments and decisions on tenure, but in the politics of publishing, an arena in which Le Guin has enjoyed the strategic success of having been selected to edit *The Norton Book of Science Fiction.* Norton's literary anthologies dominate the lucrative market of required college texts, and there had not, until now, been a Norton collection focusing on SF. The significance of this was spelled out by George Slusser in a judiciously devastating essay-review of the collection, "The Politically Correct Book of Science Fiction: Le Guin's Norton Anthology":

> Norton Anthologies, through their usefulness in the classroom, are notorious canon-makers. And because of this *The Norton Book* will always be there, offering a large number of stories at a reasonable price, it will be used in the ever-increasing number of courses offered in colleges and universities in this field. Its very economic presence then represents, *de facto,* canon. In today's academic climate however, market shares can be sizeably increased if the "correct" theoretical spin is given to a book. . . . [However, SF has proven] recalcitrant to legislation from without by any ideology, let alone the so-called ethical demands of the academy's politically correct. . . . It is not enough to season recognized SF texts (as is done here) with stories by little magazine writers and others who have little or no connection . . . with SF. SF texts must be bent to the will of the righteous. . . . The result is a masterpiece of totalitarian propaganda.[12]

I would surely not go so far as to call it a masterpiece, nor is the thrust of its propaganda totalitarian, except insofar as Le Guin excludes most writers with whom she has deeply rooted ideological differences. That is an editor's prerogative, of course, but since it would not look well for Le Guin to be seen exercising that prerogative with too procrustean a zeal when editing an anthology that is supposed to represent the genre *tout court,* she has found two ingenious ways to conform the playing field to her own preferences. Both are set forth in the collection's subtitle, *North American Science Fiction, 1960–1990.* The first geographical limitation both allows Margaret Atwood and some few other Canadians *in* and, more significant, keeps *out* all English writers, though 1960 through 1990 were the years in which British SF came into its own. Thus, there are no stories by Arthur Clarke, J. G. Ballard, Michael Moorcock, Brian

Aldiss, John Brunner, Angela Carter, Josephine Saxton, Christopher Priest, Ian Watson, or by such fellow-travelers of the New Wave as Norman Spinrad, John Sladek, or myself, even though as a publishing phenomenon British and American SF in this period were in a constant ferment of cross-fertilization.

The exclusion of SF stories written before 1960 debars an even wider host of Le Guin's ideological enemies and professional rivals. *The Norton Book* excludes, by definition, the classic short fictions of Asimov, Bradbury, Heinlein, C. M. Kornbluth, Alfred Bester, Walter M. Miller, Jr., and Kurt Vonnegut (none of whom makes the cut), and it limits the work of other writers prominent in the bonanza years of the 1950s to stories that show their imaginative powers in decline or near eclipse. A novice reader, judging only on the basis of these reprints of stories by Damon Knight, Theodore Sturgeon, James Blish, Frederik Pohl, and Robert Sheckley, would suppose these dinosaurs from the era of *Astounding* and *Galaxy* are suitable candidates for extinction along with Asimov, Bradbury, and the others who have been excluded altogether.

These exclusions facilitate Le Guin's aim to create an anthology that is a one-volume affirmative action campaign, remedying the genre's perceived historical neglect of women and other exemplary victims. Here is a partial statement of her editorial credo:

> I wish science fiction were not as white as it is. I wish I understood why it is still so white. I am glad and proud of the African American and Native American presences in this book but sorry there are not more, nor any Asian or Latin voices. . . . [13]
>
> I wish science fiction were not as male as it is, but it isn't as male as it was, not by a long shot. The strong and brilliant female presences in this book give me joy, and that so many of them are young gives me confidence. We have regendered a field that was, to begin with, practically solid testosterone.[14]

The editor herself does much to make SF "not as male as it is" and to enhance the female presences that are there not only by a policy of preferential hiring (of the the sixty-seven stories, twenty-six are by women) but also by choosing relatively feeble or ephemeral stories by older big-name male writers while representing women writers by their longer and stronger stories. Thus, ten stories by Robert Sheckley, Lewis Shiner, Damon Knight, Joe Haldeman, Barry Malzberg, Roger Zelazy, Frederik Pohl, Harlan Ellison, Robert Silverberg, and Gene Wolfe are allotted the same number of pages (forty-five) as are taken up by two stories by Suzette Haden Elgin and James Tiptree, Jr. (the pseudonym of Alice Sheldon).

Few writers are at their best writing short-shorts (as Le Guin surely knows from having herself written many ephemeral vignettes in the two- to seven-page range), but Le Guin has not even sought out the best SF by those males she has chosen to miniaturize. Rather she has chosen such stories as best mirror her own few, simple, political opinions, as set forth in the 1973 Bellingham speech noted above and repeated with little variation in her Introduction: war is wrong, and men are to blame for it; science is inhumane, and men are to blame for it; capitalism is heartless, and men are to blame for it.

In the story by Barry Malzberg that Le Guin has chosen for *The Norton Book*, the author sounds as though he is lip-synching the editor. Here is the gist of the oration for which the story serves only as a frame: "We must abandon the space program . . . because it is destroying our cities, abandoning our underprivileged, leading people toward the delusion that the conquest of space will solve their problem and it is in the hands of technicians and politicians who care not at all for the mystery, the wonder, the intricacies of the human soul. Better we should solve our problems on Earth before we go to the Moon."[15] No one, encountering such ill-written rant, would ever suspect that Malzberg can be, even in stories as short as this, mordant and funny. But those stories would not have the PC flavor that Le Guin is after.

I had my first suspicions as to Le Guin's editorial strategy before *The Norton Book* was published, since she had asked to reprint a short-short I had no wish to be remembered by. I demurred and suggested several stories, of the more than one hundred I've published, as alternatives. Brian Attebery, her coeditor, insisted that no other story but "The Apartment Next to the War" would serve their purpose. At this impasse I replied, "If I may be allowed a moment of sincere self-deprecation, I think [the story] inconsiderable and would as soon be represented by my Aunt Cecelia's recipe for lemon pudding."[16]

One of the governing assumptions of *The Norton Book*, and of feminist science fiction in general, is the equation of male sexuality with aggression and physical violence. It is usually assumed, by those of Barr's and Le Guin's way of thinking, that this does discredit to the male sex. But even among feminists, there are those who, even while they reprehend the excesses of Gorean man, rather enjoy fantasizing about women assuming the role of warriors. Joanna Russ blazed this trail in her short stories about the female mercenary, Alyx, who also stars in her 1968 novel, *Picnic on Paradise.* Alyx is a shrewd, tough daughter-of-a-bitch who kicks ass when ass has to be kicked. Her imitators have been legion.

In 1975 Russ published *The Female Man*, which must be accounted the best feminist science-fiction novel of all time, though its philosophy is utterly at odds with Le Guin's. While the politically correct Le Guin would like, in effect, to disarm men, Russ wants to empower women. *The Female Man* sprang from the acorn of Russ's 1972 Nebula-winning story, "When It Changed," which describes the reappearance of male astronauts on the colony planet of Whileaway, for thirty generations a planet populated only by women. The moment when the Whileawayan narrator gets her first sight of the opposite sex is the defining moment of feminist SF, a "first contact" on a par with the landing of the mothership in *Close Encounters:*

> They are bigger than we are. They are bigger and broader. Two were taller than me, and I am extremely tall, one meter, eighty centimeters in my bare feet. They are obviously of our species but *off*, indescribably off, and as my eyes could not and still cannot quite comprehend the lines of those alien bodies, I could not, then, bring myself to touch them, though the one who spoke Russian—what voices they have!—wanted to "shake hands," a custom from the past, I imagine. I can only say they were apes with human faces.

Little wonder that Heinlein, in *Friday*'s alphabetized dedication and tribute to twenty-nine members of the fair sex, pays his devoirs to a Jeanne, a Joan, and a Judy-Lynn, but omits the name of Joanna. Now that, truly, is a compliment.

On the whole, Russ's strategy of hypothesizing women who can cope in a "man's world" rather than, as Le Guin would have it, remodeling human nature on a maternal template, seems to have carried the day, not only in the real world, where women are being integrated into the armed services, even in combatant roles, but in the virtual reality of science fiction, where the most popular new women writers—C. J. Cherryh and Lois McMaster Bujold—have taken their cue from Russ, writing gung-ho Realpolitik space operas that make the author of Gor look like the wimp he was. Their attitude would seem to be: Feminism? Okay, we won that war. Now let's move on to the next conflict.

Sex will, of course, remain a major conflict, in the arenas of both culture and politics. While there are two sexes, one can expect them to be in contention. But sex is a conflict as well within the individual psyche. Men must always wonder what it is like to be a woman, and vice versa. And SF, because it permits such wonderings to be worked out as full-dress thought experiments, is a natural haven for those who obsess on the subject.

Russ has her Alyx, and Heinlein has his Johann Sebastian Bach Smith, a ninety-four-year-old billionaire who cheats death by being subsumed into the body (and psyche) of his twenty-eight-year-old black secretary, Eunice.[17] Le Guin, in *The Left Hand of Darkness*, invented a race of hermaphrodites, who shared, with Tiresias and Heinlein's Johann/Eunice, the special wisdom of being Both. And both John Varley (in *The Barbie Murders*, 1980) and Samuel R. Delany (in *Triton*, 1976) have posited future cultures in which elective transsexuality is a medical commonplace, on a par with cosmetic surgery. Not all the experiments have been thought experiments; science fiction surely has a higher per capita rate of transsexual surgery than any other literary genre. Four writers of significance in the field have done what Heinlein only dreamed of. The most telling such transformation was that of Hank into Jean Stine. In 1968 Hank Stine published a pornographic SF novel, *Season of the Witch*, in which a rapist is punished for his crimes by being biologically reengineered as a woman, a punishment that (as so often in pornography) becomes a source of fulfillment. That the author elected to enact that scenario in his/her own life would suggest that Stine may have been impelled, in part, by the radical feminist conviction that masculinity is a form of original sin, one from which the daughters of Eve are happily exempt.

Jessica Amanda Salmonson, another SF transsexual, having disarmed herself as man, has become an exponent of that vein of revisionist historiography that envisions ancient history as a kind of feminist utopia ruled by warrior queens. In her introduction to *Amazons!* an anthology of feminist "heroic fantasy," Salmonson declares:

> From a scholarly basis, we can show only that the jural and social systems of many early peoples were matrifocal, matrilocal and matrilinear and the position and power of women under these societies was greater than it is today. . . . It is tempting to extrapolate from . . . modern anthropoligical and available archeological evidences, an age of sophisticated worldwide matriarchal civilization.
>
> . . . We can only speculate what the common experience may have been, which produced the universal myth of women's previous rule—and in the fantasy story especially this speculation can be imaginative and entertaining, as well as, very possibily, as close to truth as any scientific hypothesis. . . . We can at least dismiss all detractors who continue, sometimes hysterically, to reason that women were never mighty. Amazons have lived and fought from the Neolithic to the streets of Chicago, Belfast, and Peking.[18]

At the risk of sounding hysterical, I would suggest that this is wish fulfillment posing as scholarship and as little to be taken seriously as similar rewritings of history by Von Daniken (who maintains that any sizable ancient artifact was the work of alien visitors), or believers in the lost continents of Mu and Atlantis, or Afrocentrists who claim that the pyramids were built by the (now lost) superscience and psychic powers of black Africans. In all these cases, the tropes of SF have been appropriated by intellectual hucksters who know that fiction can be merchandised with more success if it is repackaged as fact.

This is not to deny the psychological benefit to be derived from daydreams of heroism. If guys can have their Beowulfs and Rolands and Captain Videos, why shouldn't girls (and transsexuals) enjoy their own warrior imaginings? In fact, they do, and the paperback racks are full of space operas in which, as in Handel's operas, sopranos can lead armies into battle. And so the latest *Star Trek* spaceship is commanded by a woman—as are, increasingly, units of the armed services.

The feminist incursion into SF has produced its own distinctive icon, one that has joined the small repertory of images, like the spaceship, the robot, and the dinsosaur, that semaphore Science Fiction Spoken Here. That new icon is the Bald Woman. We saw her first in 1979 as Ilia, in the first *Star Trek* movie. Even critics who found that movie dull praised Persis Khambatta's "riveting" performance. One must suppose that those critics were riveted by her fashion image, since her acting consisted primarily of staring with dazed earnestness directly into the camera. Certainly Persis's new look made a big impression on the art director of the most influential SF magazine of that day, for beginning in 1983 *Omni* began to feature the Khambata hairdo on its covers almost as regularly as it did the starry heavens and naked eyeballs. Between 1983 and 1992, nineteen of the magazine's covers played variations on that theme. There were bald women with lucite heads, with metal heads, women whose naked skulls had metamorphosed into CD discs or conehead-style capsules. Naked female scalps split open to release various geometric figures or were decorated with globs of oil paint.

It may be that, like planets, shaved heads are easier to render with an airbrush; they are also "timeless" in a way that no actual head of hair, however styled, can be. Such an image suggests a future, "evolved" humanity that is, as Poe foresaw, all brain. It also suggests the woman as robot, an image that first made its mark in Fritz Lang's *Metropolis* of 1926, in which the heroine's evil doppelganger is first seen as an all-metal woman.

Allow all that: the real meaning, and fascination, of the icon resides elsewhere. Shaved heads on men signify naked aggression. Before De Niro goes berserk in *Taxi Driver,* he signals his intention by giving himself a mohawk. Ditto, Woody Harrelson in *Natural Born Killers.* Ditto, indeed, every Marine Corps recruit. A career of active violence is traditionally prefigured by the Haircut. Hair signifies, in almost every culture, the feminine. It is vulnerable (think of Pope's "The Rape of the Lock"); it is also a mark of vanity, and so a proper sacrifice for some higher ideal, whether inflicted as a symbolic punishment (French women who consorted with Nazis were shaved bald) or assumed as an ascetic discipline. Shannon Faulkner's last legal battle in her successful effort to enter the Citadel was over the issue of the Haircut.

As an SF icon, the Bald Woman plays off those established meanings, but its central significance is one of empowerment rather than diminishment. For all her vows of chastity, the bald Persis Khambatta enjoys a transcendental orgasm at the end of *Star Trek: The Movie*. As for Sigourney Weaver, the star of *Alien* (1979), *Aliens* (1986), and *Alien* 3 (1992), she was to become the very goddess of baldness. In the first two movies, as she confronts the most horrific, and nightmarishly phallic, alien monster ever to be seen in the movies, Weaver maintains some of the usual protocols of femininity. Indeed, her affection for a darling kitten almost costs her her life. In the third movie, she is required to make the supreme sacrifice: her head is shaved. She becomes the New Woman, who is, in the prophetic words of Joanna Russ, a Female Man.

This is not to suggest that any haircut, however drastic, will ever accomplish even a cease-fire in the war of the sexes, much less bring an end to the hostilities. The dichotomy is ageless, deep, and inescapable. Most SF readers (and writers) must cross the border between Women's Country and Man's Country on a daily basis, and each has his or her own characteristic method of dealing with the perils and inconveniences. Some simply ignore the aliens: no one (except in English class) can *make* you read Le Guin (or Heinlein, for that matter). Others—radical feminists *and* die-hard male chauvinists—never enter enemy territory without a weapon and a jealous sense of honor. There's not a lot for them to read that won't give them some degree of heartburn, but then reading is rarely a major priority. Their favorite authors are elevated to guru status, and their novels become gospel that make all other reading matter superfluous.

Then there are those—a majority, even in SF—who consort together and are able to sort out their differences with no other mediation than jokes and banter. The authors most admired by such readers can mimic each other so well in their best fictions that, without a pronoun or photo pointing the way, no reader can be sure whether the prose he or she is reading was written by him or her.

Chapter 7

WHEN YOU WISH UPON A STAR—SF AS A RELIGION

In the summer of 1964 I was invited to attend the Milford Science Fiction Writers Conference run by Damon Knight and his wife, Kate Wilhelm, at the Anchorage, their tumbledown Victorian mansion in Milford, Pennsylvania. Since its inception in 1956, the annual eight-day gathering of writers and their spouses had come to be regarded as the hottest ticket in the SF community, a sign that one was a name writer and not just another hack. Regular Milford attendees at that time included the Knights, James Blish and his wife, Virginia Kidd, Harlan Ellison, Carol and Ed Emshwiller, Joanna Russ, Keith Laumer, the anthologist Judith Merril, other editors and agents, and Richard McKenna (of *Sand Pebbles* fame). I was twenty-four and had published some dozen stories in SF magazines; to be asked to Milford was Cinderella time.

But before the ball there was an ordeal to be undergone: I had to spend a night at the Red Fox Inn, a one-time boarding house some miles out of town that was the headquarters of the Centric Foundation, an organization devoted to the maximization of human potential whose founders (and only members) were the SF writing team of Walt and Leigh Richmond. Their method of collaboration was uniquely science fictional. Walt, a laconic, Burl-Ivesish fellow, would sit with a quiet smile on his lips and telepathically project his inputs to Leigh, who would translate them into *their* prose at the typewriter. Leigh had worked at various small newspapers and come to the SF field, and to Walt, late in life by way of Dianetics, the embryonic form of Scientology. They left Scientology, or were excommunicated, just as the cash began to roll in and hierarchies gelled.

Though driven from the fold, they continued "auditing" potential recruits. Invented in 1950 by the SF writer L. Ron Hubbard, auditing is a low-brow, do-it-yourself form of psychoanalysis, in which the auditor-analyst roots up "engrams"—repressed traumas often encoded in psychosomatic ills. With a copy of Hubbard's *Dianetics* in hand, a little black box called an e-meter, and the right manner, anyone could play doctor.

Walt had the manner down pat, and no sooner had we arrived at the Red Fox than he set to work auditing my fellow lodger, Jon De Cles. Jon claimed to be suffering from a sore knee, and Walt discovered, with a few manipulations of the knee and a bit of telepathic interrogation, that Jon had had problems in his childhood with his father. Jon confirmed this, and Walt then began to ferret out the father's sins that time had been encrypted into the son's sore knee. Though encouraged to witness the entire process and have my own turn, I retired early—to a bedroom that Leigh informed me was known to be haunted by a quite unfriendly ghost. The following morning, undisturbed by the ghost but appalled by the Richmonds, I moved to a hotel for the rest of my stay in Milford.

At the conference, the Richmonds were the source of another drama when their collaborated manuscript came to be workshopped by their fellow writers. Their story was a variant of van Daniken's fancy that in the lost Atlantean past, the earth supported a civilization of extraterrestrial supermen who inseminated indigenous homonids with their own seed by way of bringing the species up to par. After the Atlanteans self-destructed, due to the "avalanche" of a solar tap, the process was renewed with scaled-down solar taps (the pyramids), and history in its usual, verifiable sense commenced. The Richmonds' reordering of world history was not without some inconsistent features that those of a skeptical temperament might choose to focus on. Many at the workshop did take such a tack, while others, though able to swallow Atlantis whole, presumed to give the Richmonds advice on the elements of style. Both forms of criticism were deeply resented, and before the Richmonds departed, Leigh explained loftily that their story, though taking the marketable form of science fiction, was, in essence, a true picture of our planet's early history and therefore exempt from the specious criticism of (in Poe's words) "mere doubters by profession—an unprofitable and disreputable tribe."

The Richmonds were the first full-throttle, off-the-wall lunatic fringers that I met in the field of SF, and they remain the purest specimens I've known. I admit that I never was face to face with those reputed to be flakiest—L. Ron Hubbard himself or A. E. van Vogt—but I have had Jacob-and-the-angel sessions with Theodore Sturgeon, Philip Dick, and the film director Alexander "El Topo" Jodorowski, all of whom were experts in the art of spiritual arm wrestling and quite unwilling to accept anything less than unconditional surrender. What each of these writers had in common, Walt and Leigh included, was a deeply rooted conviction of his or her own genius—not simply in a literary sense but as a direct psychic connection to some Higher Wisdom. A number of SF writers have commanded their own small legion of true believers: L. Ron Hubbard, most conspicuously, but also Robert Heinlein, Theodore Sturgeon, Phil Dick, J. G. Ballard, and Frank Herbert. Genius would seem almost an occupational hazard of being an SF writer, or one of the perks—or maybe, simply, a duty.

The myth of genius is, as much as rocket ships and alien monsters, one of science fiction's favored themes, its own special version of the Cinderella story. Among the first books of SF I read when I was twelve was an anthology edited by William Tenn, *Children of Wonder*, a book that should be required reading for any child thought to be wanting in self-esteem. Among its many parables of the secret wisdom (and/or psychic powers) of the prepubescent was a novella, "Baby Is Three," by Sturgeon, an update of Tom Swift for future English majors. After precocious literary successes, the young genius who is the hero of that tale becomes *More Than Human* (the title of the book version) by getting in sync with other kids with complementary psychic powers.

Walt and Leigh, with their Centric Foundation, secret Atlantean wisdom, and telepathic writing powers, might have been living in a never-never land scripted by L. Ron Hubbard, Ignatius Donnelly, Sturgeon, and assorted other SF writers, but despite appearances, theirs was no folie à deux. It was a miniature version of a unique social institution, science-fiction fandom. SF fandom sprang into being in the United States in the 1920s, along with the first pulp magazines, and grew steadily, until, by the 1960s there were estimated to be some ten thousand dedicated fans who kept in touch by means of mimeographed fanzines and by meeting at annual (or more frequent) conventions throughout the country.

Though a devoted reader of SF, I did not become acquainted with the world of fandom until I'd begun publishing my own SF stories in the early '60s, when I began going to conventions to meet the writers whose work I'd been reading for years, to network with editors, and to promote my own books. For established writers, SF conventions provide holidays with large dollops of egoboo (ego-boost, a fannish coinage); for aspirants they served (along with the fanzines) as a bridge from amateur to professional status. Since at least the '60s a majority of writers, editors, and agents in the SF field have come up through the ranks via fandom.

But fandom is only peripherally a system of apprenticeship. Fandom is a way of life. Indeed, there is another fannish coinage, *fiawol,* that is simply an acronym of that holy truth. As a way of life, fandom offers many of the benefits of disorganized religion—religion, that is to say, of the New Age variety, emphasizing self-fulfillment at the expense of doctrinal orthodoxy or a code of ethics, without tithing, pricey churches, or official hierarchies, but a religion even so, whose gospel is preached at conventions and in fanzines.

The first tenet of fandom is that SF is the true and only literature. "Mundane" fiction, which professes to mirror the real world, is a deceitful heresy; SF is visionary, a map of the future by means of which fans have a private view of the millennium—which fans shall inherit. In SF's infancy, this gospel existed mostly in an uncodified form in the hearts of the faithful, but it became official doctrine in the work of the critics Alexei and Cory Panshin, whose fannish manifesto, *SF in Dimension* (1976), proclaimed that "mimetic fiction"—all fiction that is not SF—is "a negative drag of literature" and that even "SF which rejects its freedom to be positive is as big a bummer as mimetic fiction." SF's mission is nothing less than to present a vision of the "future selves imminent within us," when men shall transcend their mortal coils and be as gods. Outside of science fiction there is no salvation—not simply in a literary sense but with respect to the Fate of All Mankind.

The second tenet of fannish faith is that fans are a breed apart, elevated above the uncircumsized by a mysterious, inherent difference. For the young (as most fans are) this will present no difficulty. Any teenager with a modicum of self-esteem knows this of himself already. For those who have been relegated to the category of nerd in high school years, however, there can be comfort in the assurance that despite being members of the chess club rather than the football team, they will have the last laugh—as indeed many will, as they parlay their brains into scholarships to MIT or similar elite institutions.

Yet there will be those, like (I imagine) Walt Richmond, whose capabilities don't gibe with their aspirations, whose chess game isn't top-notch and whose grades, even with effort, are C's and B's. How is one to reconcile, in such cases, the discrepancy between a grandiose self-image and the steady encroachments of mundane reality?

The usual answer has been religion in one form or another. Resign yourself to less than you'd been counting on in the way of earthly success and lay up treasures in heaven, where your innate excellence will finally be recognized. Meanwhile, you may enjoy the companionship of those other elect souls who share your earthly fate and your blessed destiny.

In its ideal form (say, as a Carthusian monk) such a philosophy may promote a saintly existence, but in practice it often leads to a vindictive resentment of those outside one's own small fold and to daydreams of millenarian revenge, when the legions of the anti-Christ will be incinerated by one's ally on high. Recently, this malignant form of millennialist, quasi-religious SF reached its apotheosis in the outrages perpetrated by Japan's Aum Shinrikyo cult, when members released the nerve gas sarin in Tokyo's subways on March 20, 1995.

Aum Supreme Truth (as it calls itself in English) was the inspiration of a half-blind, self-made guru, Shoko Asahara, who parlayed a shopfront operation peddling yoga lessons and herbal remedies into a multimillion dollar cult with plans for taking over Japan and then the rest of the world. Though Aum's plans were thwarted, it did manage in only a few years to build its own secret factories for the manufacture of arms and biological weapons. Before they deployed sarin, there were earlier failed experiments with botulinus and anthrax, as well as long-term aspirations to obtain nuclear capabilities. Because Aum recruited intensively from Japan's best universities, its ambitions came close to being realizable. Aum had the technological know-how, the economic resources, and a sincere desire to bring about the end of the world, or as much of it, at least, as Aum could lay waste to.

Why would Aum want to pursue such a course? To say that Aum's rank and file were under the influence of a demented, embittered, charismatic guru begs the question. The upper echelons of the organization fully shared Asahara's vision and his zeal for achieving Apocalypse now. Even after the original attack, when the entire leadership was in jail, cultists continued to attempt further mass murders in the Tokyo subways. It was no longer a means to an end; it was the goal.

Aum's minions were the children of Godzilla. They had grown up watching cartoons like *Space Battleship Yamato,* in which half-human cyborgs wreak awe-inspiring devastation on whole cities. They'd graduated to the *gegika,* book-length comics featuring gung-ho tales of rape and murder against *Bladerunner*-esque backdrops. It was in the *gegikas* and in the Japanese *Twilight Zone* (a magazine specializing in New Age wonders, including a photo spread of the guru himself "levitating" in lotus position) that Aum advertised Asahara's books, such as *Secrets of Developing Your Supernatural Powers* and *Declaring Myself the Christ.* The ads for the earlier, more modestly titled book declared, "Spiritual training that doesn't lead to supernatural powers is hogwash! The Venerable Master will show you the secrets of his amazing mystic powers. See the future, read people's minds, make your wishes come true, X-ray vision, levitation, trips to the fourth dimension, hear the voice of God and more. It will change your life!"

Once Aum got hold of a recruit, he or she was subjected to the entire repertory of cult brainwashing techniques: social isolation, starvation, sleep deprivation, and drugs. "Monks" and "nuns" proved their commitment by deeding over all their worldly goods to Aum. To synchronize their brain waves with those of the guru, they were fitted with "electrode caps" called the PSI, or Perfect Salvation Initiation: snug battery-powered hoods that delivered six-volt shocks to their scalps at regular intervals. At its height, Aum had little factories dotted across Japan staffed with thousands of workers in PSI helmets busily assembling machine guns, manufacturing nerve gas, baking cookies impressed with the Aum insignia, and sewing ceremonial robes. Welcome to the fourth dimension.

Aum's theology was a syncretistic mishmash of Hinduism, Tibetan Buddhism, and its own special variety of Christian millennialism, in which Aum and Asahara would reign supreme after the Day of Judgment. Only the upper echelon knew that that day was to be hastened by Aum's own genocidal initiatives, but as in Nazi Germany, the rank and file had good reason to suspect that their leaders were preparing a Final Solution. Asahara's writings, like *Mein Kampf*, resonate with apocalyptic menace.

Science fiction does not have a copyright on Armageddon and the Apocalypse. Those notions have been in perennial bloom since the time of Ezekiel and St. John. Indeed, Aum's whole arsenal of wonders—from UFOs ("UFOs often appear on Earth these days," Asahara explained in a 1990 lecture. "It will become one of the main factors of Armageddon whether we can benefit from UFOs or not.") to ESP—may be considered part of pop culture and no longer specifically SF. But Aum does have a specific SF connection in the work of Isaac Asimov, whose *Foundation* series provided a crucial element of the Aum mythology. In Aum's version, Asahara takes on the role of Asimov's Hari Seldon, a genius who discovers the laws of "psychohistory," which predicts, infallibly, that "interstellar wars will be endless. Interstellar trade will decay; population will decline; worlds will lose touch with the main body of the Galaxy." The answer to this threat is a secret society of subsidiary geniuses to act as guardians of civilization's flame during the destined dark ages.

"The similarities [of Asimov's *Foundation*] to Aum and its guru's quest were remarkable," note David Kaplan and Andrew Marshall, in an authoritative history of the cult.[1] "In an interview, Murai [one of Aum's inner circle] would state matter-of-factly that Aum was using the *Foundation* series as the blueprint for the cult's long-term plans. He gave the impression of 'a graduate student who had read too many science fiction novels,' remembered one reporter. But it was real enough to the cult. Shoko Asahara, the blind and bearded guru from Japan, had become Hari Seldon; and Aum Supreme Truth was the Foundation."

Asimov cannot be blamed for Shoko Asahara's megalomania any more than Heinlein can be held responsible for the deeds of one of his many admirers, Charles Manson. A vivid representation of the abuses of power, if it reaches a wide enough audience, will inspire some part of that audience to daydreams of attempting the same crimes. Novels about serial killers are undoubtedly the favorite reading material of potential serial killers, but the link between daydream and crime need not be so direct, as witness Aum's expropriation of Asimov's work. Among the SF writers of his generation, few others have a more consistent track record than Asimov for upholding liberal causes and championing common sense against New Age charlatanry.

Science fiction is in its nature millennialist. All through the twentieth century, it had its eye fixed on 2001 and beyond. It revels in global catastrophes. My own first novel, *The Genocides* (1965), chronicled the extinction of the human race, and J. G. Ballard brought the world to an end four different ways in his first four novels. Wells was there before us, of course, and John Wyndham, famed for *The Day of the Triffids*. Although the Almighty is usually not directly implicated in these SF catastrophes, an undercurrent of apocalyptic justice does color such proceedings, often of an antitechnological bent, so that humanity is hoist by its own high-tech petard. In Kurt Vonnegut's best SF novel, *Cat's Cradle* (1963), the agent of doom is a human invention gone awry, a chemical agent that causes all water to freeze irreversibly. In a memorable story by James Tiptree, Jr. (aka Alice Sheldon), "The Last Flight of Dr. Ain," an immunologist circles the globe by jet, spreading a mutated leukemia virus that will render the human race extinct. His motive is to save the woman he loves, who is none other than Gaea, Mother Earth herself, goddess of the New Age, who can be brought back to health only by this supreme sacrifice. One can easily imagine Tiptree's story, or Le Guin's mammoth antitechnology tract, *Always Coming Home* (1985), inspiring another cult to take the path of Aum Shinrikyo, committing mass murder not for the glory of Asahara but for the greater ecological good. So-called "deep ecologists" (radical leftists who advocate the revocation of the Industrial Revolution and a worldwide return to Stone Age population levels) have already found such a benefit in the plague of AIDS.

Fortunately, a hunger for divine retribution is not the dominant element of most religious faiths. Fear of death and mortal illness usually bulk much larger. Religion's traditional answer to this has been the promise of an afterlife in which such intolerable truths will be transcended. This has also been the answer of those like Alexei Panshin, for whom SF serves as a religion, a religion that can be traced back to Poe's "Mesmeric Revelation" and "The Facts in the Case of M. Valdemar."

Poe was too devoted to his literary fame to undertake the hard work of founding a new religion, but that task was undertaken, not long after his demise, by two extraordinary women. One of them has already been noted in Chapter 2—Madame Blavatsky, founder of the Theosophical Society, which still survives as the eldest of New Age clubhouses. The other was Mary Baker Eddy, who founded Christian Science.

It would be more precise to say that Mrs. Eddy found Christian Science rather than founded it, for it was the invention of a New England mesmerist-turned-faith-healer, Phineas Parkhurst Quimby. Mrs. Eddy, an imaginary invalid of indestructible strength, first was Quimby's patient, then his student, and finally his usurper and plagiarist, who lifted her magnum opus, *Science and Health*, from his own radically gnostic writings. According to Quimby (and Mrs. Eddy), ill health is only a mistaken idea. Illness, evil, death, and even poverty exist only in the mind, and if one will only rid one's mind of such ideas, the Goodness of Everything will become manifest. This can be a very liberating gospel to anyone in sturdy health and reasonably well off, for it frees one of any responsibility to those less fortunate except to lecture them on the principles of Christian Science.

Mrs. Eddy soon set herself up in the lecturing business and let her subordinates carry on the work of healing, for which she seemed to have an imperfect knack. She was, however, a dynamite franchiser, who oversaw her expanding empire with the jealousy of Jehovah and the merchandising skills of Walt Disney. At one point she sought to hire a hit man to get rid of a subaltern who had begun to steal her thunder. His "malicious mesmerism," exercised at a remote distance, provoked her to outrageous libels and threats against his life and, finally, an actual attempt, which led to a grand jury indictment of her husband and another follower.

The scandals surrounding Mrs. Eddy and the establishment of Christian Science continued to the end of her life in 1910, at age ninety. Her biography was the first book written by one of America's most distinguished writers, Willa Cather.[2] Mrs. Eddy's followers did all they could to suppress its publication, and when that effort failed, they bought up all the copies they could and checked it out of libraries, permanently.

Further, "the copyright . . . was purchased by a friend of Christian Science, the plates from which the book was printed were destroyed, and the original manuscript also acquired." It is little wonder that the book has been unavailable through most of this century.

Most New Age prophets and profiteers derive whatever systematic theology they have to the scheme set forth in *Science and Health:* Only the spiritual world is real; the physical world, including our mortal bodies, is an illusion. Those who can see through that illusion will enjoy good health, immortal life, and Porsches. The most enduringly successful of Christian Science–descended religions in our time has been Scientology, the brainchild of the hack writer L. Ron Hubbard. As an SF writer, Hubbard had a neglible impact on the field, and it is doubtful that without the controversies surrounding Scientology and its immense success, his fiction would be read or remembered today. *The Science Fiction Encyclopedia*'s estimate of the merits of Hubbard's early SF, which mostly appeared pseudonymously in the more downscale pulp magazines of the 1940s, is not the stuff a writer's dreams are made of:

> His best-known early SF novel, *Final Blackout* (1940), grimly describes a world devastated by many wars in which a young army officer becomes dictator of the UK, which he organizes to fend off a decadent USA. It cannot be denied that the book veers extremely close to the fascism its text explicitly disavows. But SF was clearly not Hubbard's forte, and most of his work in the genre reads as tendentious or labored or both. . . . In general his early work, though composed with delirious speed, often came to haunt his readership, and its canny utilization of Superman protagonists came to tantalize them with visions of transcendental power.
>
> The vulnerability of the SF community—from Campbell [editor of *Astounding Science Fiction*] and A. E. Van Vogt down to the naivest teenage fans—to this lure of transcendence may help account for the otherwise puzzling success first of Dianetics, then of Scientology itself.[3]

Few SF critics have had kinder words for Hubbard's SF, with the exception of Algis Budrys, but Budrys's encomiums cannot be taken at face value, having been written only after he had become, in effect, an employee of Scientology, as a director of its Writers of the Future program.

Hubbard may have been a writer of less-than-genius level, but he was a shrewd, clear-headed, and capable con man. Foreseeing that his destiny as an SF writer would not take him into the first rank, Hubbard shifted course sometime in 1949 and began to concoct the "science" of Dianetics. Never one to conceal his light beneath a bushel, he also boasted at this time, to a group of science-fiction fans hosted by Sam Moskowitz, "Writing for a penny a word is ridiculous. If a man really wanted to make a million dollars, the best way to do it would be to start his own religion."

In 1950, using the May issue of Campbell's *Astounding Science Fiction* as his launchpad, Hubbard started his new religion, with the wholehearted endorsement of editor Campbell: "I want to assure every reader, most positively and unequivocally that this article is *not* a hoax, joke, or anything but a direct, clear statement of a totally new scientific thesis."[4] In his introduction to the book version, *Dianetics*, which appeared at the same time as the article in *Astounding*, Campbell waxed still more eloquent: "The creation of Dianetics is a milestone for Man comparable to his discovery of fire and superior to his invention of the wheel and the arch. . . . The hidden source of all psychosomatic ills and human aberration has been discovered and skills have been developed for their invariable cure."[5] (Campbell was speaking from experience in this instance: Hubbard's ministrations had much improved the sinusitis that had plagued him for years.)

The book took off. Piqued by Campbell's hype and encouraged by the endorsement of a genuine physician, Dr. Joseph Winter, a general practitioner from St. Joseph, Michigan, thousands of SF fans bought copies and began to "audit" their friends, who in turn had to do the same thing. The process was essentially the same as I'd witnessed between Walt Richmond and Jon De Cles. Russell Miller, the author of a muck-rich biography of Hubbard, describes the modus operandi, in its unevolved 1950 version:

> Auditing began in a darkened room by inducing in the pre-clear [the analysand] a condition Hubbard described as a "Dianetic reverie," which could apparently be recognized by a fluttering of the closed

> eyelids. It was not so much a hypnotic trance, he was careful to point out, as a state of relaxation conducive to travelling back along the time-track. Once the reverie had been induced, the auditor placed the pre-clear back in various periods of his life, moving inexorably towards birth or conception. Most pre-clears, Hubbard advised, would eventually experience a "sperm-drive" during which, as an egg [*sic*], they would swim up a channel to meet the sperm [*sic*]. Once the earliest engram had been erased, later engrams would erase more easily.
>
> An average auditing session would last about two hours and Hubbard estimated that a minimum of twenty hours' auditing would be needed before the pre-clear began to reap the rewards.[6]

As a parody of orthodox Freudian doctrine and practice, *Dianetics* is bang-on, but that, of course, was not its intention. Rather, it was a generic, over-the-counter version of psychoanalysis available, initially, for a fraction of the price. In lieu of oedipal fantasies, it offered simpler ways to lay the blame on dad and, preferably, mom:

> If a husband beat his pregnant wife, for example, yelling, "Take that! Take it, I tell you. You've got to take it," it was possible the child would interpret these words literally in later life and become a thief. . . .
>
> Some of the worst pre-natal engrams were caused by naming the child after the father. If the expectant mother was committing adultery, as many of Hubbard's pregnant women were wont to do, she was likely to make derogatory remarks about her husband while engaged with sexual intercourse with her lover. The foetus, obviously, would be "listening" and if he was given the husband's name he would assume in later life that all the horrible things his mother had said about his father were actually about *him*.[7]

With these and many other such hints from the prompter's box, novice auditors and pre-clears (the cult's name for paying recruits) knew what tales to tell. Jon De Cles had a sore knee? It was his father's fault. One need not even be a Scientologist to reap the benefits of such therapy, for it does not differ in essentials from the "hypnotic regression" therapy by which resentful daughters may discover themselves victims of Satanic sexual abuse or UFO abductees may ferret out the truth about their "lost hours." *Dianetics* represents the opening of a huge Pandora's box of specious or unprovable grievances.

Auditing did not remain for long a low-cost alternative to psychoanalysis. Soon after the book appeared, the Hubbard Dianetic Research Foundation opened a training course for auditors. The $500 price tag was a comparative bargain compared to what later initiates would have to pay. Publicity brought controversy, with debunkings in *Scientific American* and major newspapers, but that publicity only brought in new converts. The adverse reactions also helped accelerate the process that Hubbard initially had in mind: Dianetics, the science, morphed into Scientology, the religion. This had many advantages. Religions cannot be taxed; even better, they are exempt from criticism or hostile comment—at least in the United States.

This transformation, begun in 1952, involved an upping of the ante with regard to Dianetics doctrine, an evolutionary process in which traditional SF tropes became ever more bizarre, trumping the simplified Freudianism. It turns out that we are *all* descended from extraterrestrials! Indeed, by virtue of reincarnation, we have lived numerous lives as "thetans":

> In existence before the beginning of time, thetans picked up and discarded millions of bodies over trillions of years. They concocted the universe for their own amusement but in the process became so enmeshed in it that they came to believe that they were nothing more than the bodies they inhabited. The aim of Scientology was to restore the thetan's original capacities to the level, once again, of an "operating thetan" or an "OT." It was an exalted state, not yet known on earth, Hubbard wrote. "Neither Lord Buddha nor Jesus Christ were OTs according to the evidence. They were just a shade above Clear."[8]

For the three decades from 1952 to 1982, Hubbard took a holiday from hack writing and concentrated on the serious work of organizing his religion. Scientology filled its coffers by merchandising various levels of auditing and workshops to the "raw meat," as Hubbard styled the church's marks. As in Aum, the more vulnerable recruits could find their life savings wiped out by a few weeks of wisdom. Also, as in Christian

Science and Aum, the church diversified its activities into areas like financial consultancy, health spas, and drug treatment programs—fronts that concealed their Scientological origins. In its May 6, 1991, issue, *Time* magazine devoted a cover story to an exposé of Scientology's long rap sheet of scams and felonies. Scientology would seem to be motivated, like its founder, chiefly by greed and has no ambition to hasten the Day of Judgment or to stockpile nerve gas—a cheering thought, given the size the organization has grown to (by *Time*'s estimate, 50,000 active members; by the church's, 8 million).

Although there is no denying (except by Scientologists) that the organization's history is one long cascade of scandals, the law has been loath to deny Scientology the polite inattention that it traditionally affords to all but the most egregious of religious con men, such as Jim and Tammy Bakker. Indeed, much of Scientology's bad repution derives from behavior that is common practice among religious zealots. It was Christ who commended his followers to leave their families and follow him, and Jim and Tammy have not been alone in going after not just the widow's mite but the deed to her house as well. These are immemorial religious traditions, and so it's little wonder that the National Council of Churches and the Catholic church have entered *amicus* briefs in support of Scientology in many court battles.

In 1983, after a long absence from SF, L. Ron Hubbard published the largest dime novel of all time, *Battlefield Earth*. Nowhere in its press release was the word *Scientology* used, and the book itself does not overtly seek to convert readers to the faith. One must suppose that Hubbard wrote it—and the ten-volume *Mission Earth* dekalogy that followed from 1985 to 1987—for art's sake much more than for profit. The cult exerted itself in promoting the books, reportedly dispatching the faithful to buy mass quantities to boost sales figures to best-seller levels—a promotional effort that other authors, lacking Hubbard's resources, can only envy, especially since most of the *Mission Earth* books appeared posthumously. Because of the immense combined length of these tomes and the author's advanced age at the time he is claimed to have written them (he was seventy-two when *Battlefield Earth* appeared, and there was even then speculation, because of his Howard Hughes–like reclusiveness, whether Hubbard was still even alive), some skeptics have suggested that this later SF was the work of an atelier rather than the author's own autumnal bloom. As a believer that genius is induplicable, even inverse genius, I am inclined to believe they are his. Surely the introduction to *Battlefield Earth,* with its Ace Ventura vanity and its instinct for *le mot injuste,* could have been written by no one else. He proclaims the book's vaulting ambition by boasting that it contains "practically every type of story there is—detective, spy, adventure, western, love, air war, you name it." And: "I had, myself, somewhat of a science background, had done some pioneer work in rockets and liquid gases, but I was studying the branches of man's past knowledge at that time to see whether he had ever come up with anything valid." And . . . but Hubbard is like Shakespeare; one could go on quoting forever.

Hubbard's main charm as a con man can be glimpsed in the passage above, where he modestly admits to having "somewhat of a science background." He was also a war hero "flown home in the late spring of 1942 in the Secretary of State's private plane as the first US returned casualty from the Far East"; the conductor, at age twenty-one, of the first mineralogical survey of Puerto Rico; a wanderer, at age fourteen, in the Far East, where "I remember one time learning Igoroti, an Eastern primitive language, in a single night." In short, anything anyone could do, *he* could do better. Or, in the more succinct words of a California judge in a 1984 case in which the church sued a biographical researcher, Hubbard was "a pathological liar."

One might never trust such a person, but one might very well be entertained by him, like those friends of Ollie North who looked upon his lies as a function of the twinkle in his eye. Many writers (the poet James Dickey, for instance) are notable romancers, careless with the truth insofar as it relates to their own lives, and they are indulged in this foible by those who consider mendacity a natural correlative of creativity. Whether the founders of religions should be accorded equal latitude is a judgment call that each believer must make for himself. Paradoxically, true believers seem to have a natural immune reaction to unwelcome truths. Scientologists have been as fiercely protective of Hubbard's posthumous reputation as have been Mrs. Eddy's followers and the Mormons as well. Perhaps the greatest strength of any cult is the unwillingness of those whom it has taken in to admit they have been deluded, cheated, and made fools of. While they still have funds to litigate against those who've exposed their frauds and other misbehaviors, they will do that, and when their backs are really against the wall, mass suicide may seem the best option, as it did to the faithful in Guyana and Waco.

With a little more help from his friends and a little less intellectual integrity, Philip K. Dick might have been the L. Ron Hubbard of the 1980s. It is greatly to his credit that he finally declined to wrap himself in the saffron robes of guru-dom, but it was a near thing.

His temptation began in February and March 1974. At that time, he was in recovery from amphetamine abuse and married to his fifth wife, the twenty-year-old Tessa Busby, who would give him the one thing he asked for in a spouse: her infinite attention. He was writing again after a long hiatus, and solvent. The future looked bright—and then the visions began to come, and they were far brighter.

First, there was the mysterious dark-haired girl who appeared at his door to deliver the Darvon he'd been prescribed after the removal of a wisdom tooth. He was fascinated by her necklace with its image of a fish, a sign used by the early Christians. In the thousand-page *Exegesis* he would soon after begin writing as an Epistle to Himself, he began to recall his earlier lives:

> When I saw the Golden Fish sign in 2-74 I remembered the world of *Acts*—I remembered it to be *my* real time & place. So I am (esse/sum) Simon reborn—& not in 2-74 or 3-74 but all my life. I must face it: I am Simon but had amnesia, but then in 2-74 experienced anamnesis [a Platonic term for the experience of recollecting eternal truths]. I Simon am immortal & Simon is the basis for the Faust legend.[9]

He was also Thomas, a first-century Christian tortured by the Romans, and Firebright, whose nature was divine. He began to dream whole museums of modern paintings, "hundreds of thousands of them," done in the style of Kandinsky, Klee, and Picasso. "I spent over eight hours enjoying one of the most beautiful and exciting and moving sights I've ever seen, conscious that it was a miracle. . . . *I* was not the author of these graphics. The number alone proved that."[10]

Dick would insist repeatedly that the content of his dreams was proof of an origin outside his own psyche, and he would give the entity communicating with him the name of Valis, an acronym for Vast Active Living Intelligence System—a Logos with a bad case of logorrhea:

> Almost each night, during sleep I was receiving information in the form of print-outs: words and sentences, letters and names and numbers—sometimes whole pages, sometimes in the form of writing paper and holographic writing, sometimes, oddly in the form of a baby's cereal box . . . finally galley proofs held up for me to read which I was told in my dream "contained prophecies about the future," and during the last two weeks a huge book, again and again, with page after page of printed lines.[11]

While Dick was thrilled to be the recipient of a constant flow of divine visitations, both blissful and sinister (like his own SF in that regard), he remained anchored enough in reality to be able to predict the public reaction once he went public with Valis's revelations: "Took drugs, saw God. BFD [big fucking deal]."

He was not that far off the mark. I visited Phil Dick for the first and only time in September 1974, when he was still in the grip of his revelations. They now included a good deal of the sort of Greek one encounters reading modern theology and a small miracle: Valis had diagnosed his young son's hernia. For twelve hours, from noon to midnight, Dick monologued, occasionally demanding of Tessa corroboration of his wilder whoppers. My memory of the occasion, as reported in an interview with Sutin, is quoted in his biography of Dick:

> We talked about whether the dreams were from an external source. How else, he wanted to know, could he have heard ancient Greek? I suggested that the part of the mind used in dreams is unlikely to know that what we're hearing is actually Greek. He didn't fancy my argument.
>
> I was fascinated. He was determined to make me concede that he'd had a religious experience, a true vision. It was like arm-wrestling for hours, and neither of us got the other's arm down on the table.
>
> I was being politely skeptical and affirming the imaginative side of his experience. At the same time I was thinking: this is a masterful con. Dick is a professional entertainer of beliefs—and what else is a con-man. He wants to turn anything he imagines into a system. And

> there's his delight in making people believe—he *loved* to make you believe. That made for great novels, but when he overdid it could become delusions of reference. The urge to translate every imagined thing into a belief or suspended disbelief is a bit of a jump. Yet it was probably Dick's ability to sew those things together that was his main strength as a novelist.[12]

Another visitor to the Dick household, Hampton Francher, a producer interested in securing rights to *Do Androids Dream of Electric Sheep?* offered a somewhat more acerbic impression of Phil on his road to Damascus:

> He was baronial, expansive, wall-to-wall effusiveness—come into my home, make dinner, "my dear," and so on. Germanic almost. There was no room for two-way conversations, no room for another ego in the same room. He was a brilliant guy. His eyes twinkled, he was congenial.
>
> I began to think he was a little . . . that he lied. He told me things which I didn't know if he believed or not. Some of these things, if he *did* believe them, I thought that he might be, not clinically, but a touch paranoid. That the FBI is after him. And he would dramatize with physical, facial characteristics that were a little "over the top," as they say in acting.[13]

If Dick was not experiencing divine revelations on a daily basis, was he then simply making it all up? To a degree, I think so. He was proud of his persuasive powers and would tailor each new account of the Valis experience to suit the expectations and vocabulary of his audience. Many of the details of our long confabulation have appeared in other reports in another, significantly different form. But that he was in an immense and happy turmoil well beyond the joy of storytelling can scarcely be doubted. Dick's biographer, Lawrence Sutin, offers a simple medical diagnosis that would account for most of the symptoms: temporal lobe epilepsy.[14] He quotes from a 1982 issue of the *Annals of Neurology*: "Such 'psychic' or 'experiential' phenomena activated by epileptic discharge arising in the temporal lobe may occur as complex . . . auditory-visual hallucinations . . . that to the affected patients assume an astonishingly vivid immediacy."[15] Even more to the point is this description of typical TSE symptomology:

> *Hypergraphia* is an obsessional phenomenon manifested by writing extensive notes and diaries. . . . Suspiciousness may extend to paranoia, and a sense of helplessness may lead to passive dependency. . . . Religious beliefs not only are intense, but may also be associated with elaborate theological or cosmological theories. Patients may believe that they have special divine guidance.[16]

Many sections of Dick's *Exegesis*, the more-than-2-million-word spiritual diary he began to keep after February 1974, would certainly qualify as hypergraphia. Take this passage, written in September 1978:

> The Savior woke me temporarily, & temporarily I remembered my true nature & task, through the saving gnosis, but I must be silent, because of the true, secret, transtemporal early Christians at work, hidden among us as ordinary humans. I briefly became one of them, Siddhartha himself (the Buddha or enlightened one), *but I must never assert or claim this.* The true buddhas are always silent, those to whom dibbu cakha has been granted.[17]

The difference in this case is that the hypergraphia was the affliction of a writer of already established reputation and uncommon gifts. True buddhas may have to keep quiet about their secret identity, but traditionally they are allowed to speak in parables. So all the while Dick scribbled away at his *Exegesis*, he also wrought from those same materials his last four novels, including *Valis*, a work of SF unlike any other, a memoir of madness recollected in a state of borderline lucidity. The two protagonists are Philip K. Dick (one of the best drawn of the many self-portraits in his novels) and Horselover Fat (the same name, rendered from Greek and German). What is most artful and confounding about *Valis* is the way the line between Dick and Fat shifts and wavers, Dick representing the professional writer who understands that all these mystic revelations are his own novelistic imaginings, while Fat is the part of him that receives, for a while, and believes, a little longer, messages from the Vast Active Living Intelligence System beaming pink laser-beam messages to him from an orbiting satellite. All of Dick's pet obsessions—Wagner, Ikhnaton, UFOs, the Roman Empire (still, like Hitler, alive and well), Richard Nixon—are conflated into one thick Jungian stew.

Valis represents Dick's second attempt to come to terms with the 2–74 experiences. The first *Radio Free Albemuth*, written in 1976 but published only posthumously, employed the same basic premise of Dick's interactions with a visionary doppelganger, but it somehow doesn't gel. Dick must have known this, because when his editor at Bantam asked for revisions, he sat on the manuscript for two years and then, in a two-week burst of glory, brought it off, writing an entirely new novel. The critical response was not unanimous, but those who did not find it off-the-rails or (worse than that) difficult were blown away. *The Encyclopedia of Science Fiction* judged it to be "the finest book of Dick's last years, a fragile but deeply valiant self-analysis," and that is some of the faintest praise it has received. Dick's biographer, Sutin, ranks it, with *The Three Stigmata of Palmer Eldritch*, as his greatest work, "a breviary of the spiritual life in America, where the path to God lies through scattered pop-trash clues. Unsanitized by sanctity, loopy as a long night's rave, it breaks the dreary chains of dogma to leave us, if not enlightened, freely roaming."[18]

But for some readers even that is too tepid a response. Inevitably, those of Dick's admirers who were as God hungry as he found in *Valis*, and in those parts of the *Exegesis* that reached print, a confirmation of their own spiritual adventuring on the fringes of New Age religion. They believed, though often they could not say exactly what they believed, for Dick's revelations were often self-cancelling. The editor of *Gnosis*, Jay Kinney, wrote in an article about Dick that was featured in the first issue of that magazine:

> Dick remained firm in his conviction that *something* had happened to him, something with a significance reaching beyond his own psyche. It was fortuitous [*sic*] for Dick that his career as a science fiction author placed him in one of the few niches in Western society where wild visions of alternate realities are accepted and honored.
>
> Myths don't need to be *literally* true to be both meaningful and use-

ful. The legend of the Holy Grail has a symbolic value quite apart from the historical question of whether there ever was an actual Grail.[19]

If that seems temporizing (*something* happened, but maybe not *literally*), then it is temporizing in a tradition that embraces revelations of far great antiquity than those of 2-74 and Dick's *Exegesis.* For centuries, Christian exegetists have been coping with the same problem and coming up with the same equivocal answers. Here, for instance, is how Joseph Campbell glosses Matthew's account of the Transfiguration (Matthew 17:1–9):

> Individual destiny is not the motive and theme of this vision, for the revelation was beheld by three witnesses, not one: It cannot be satisfactorily elucidated simply in psychological terms. Of course, it may be dismissed. We may doubt whether such a scene ever actually took place. But that would not help us any; for we are concerned, at present with problems of symbolism, not of historicity. We do not particularly care whether Rip van Winkle, Kamar-al-Zaman or Jesus Christ ever actually lived. Their *stories* are what concern us; and these stories are so widely distributed over the world—attached to various heroes in various lands—that the question of whether this or that local carrier of the universal theme may or may not have been a historical, living man can be of only secondary moment. The stressing of the historical element will lead to confusion; it will simply obfuscate the picture message.[20]

If Jesus Christ is to be allowed such latitude, if it is the *story* that should concern us, why shouldn't Phil Dick enjoy the benefit of the same doubt?

Of course, the Beatles claimed a similar coequal celebrity status with Jesus—at the time when their stories were "so widely distributed over the world." The only test of timelessness is the test of time, and by that standard Dick has been doing pretty well. Since his death in 1982, almost his entire oeuvre has returned to print. Two hit movies have been made from his novels; others are in production. There have been two biographies and continuous critical attention. Phil's Number One Fan, Paul Williams, established a Philip K. Dick Society, and for years its newsletter has documented the growth of his posthumous reputation.

It may well be that his reputation was one of those, like Sylvia Plath's or James Dean's, that benefits from a timely demise. Jay Kinney, in his *Gnosis* article, declared, "What ultimately distinguishes madness from mysticism is the direction the affected individual's life takes. For the insane, the experience leads to further disintegration; for the mystic, it leads to unification and healing."[21] By this test, Dick's batting average was about .500. In the aftermath of 2-74, he broke up with his fifth wife and their child and returned to his usual pattern of infatuated philandering, and in most other ways the resemblance to Falstaff was always closer than to Buddha. On the other hand, his writing certainly spoke of "unification and healing." His last novel, *The Transmigration of Timothy Archer*, published in the year he died, represents his single artistic success outside genre fiction. Conflating aspects of his own life with that of his friend James Pike, Episcopal bishop and celebrity author, *Transmigration* shows what Dick might have done if he'd tried to be a serious mainstream writer. He might have been quite good.

And he might have gone on to be even better. But I doubt it. Once he'd mined the materials of the *Exegesis* (a process still continuing in *Transmigration*), there would have been a slacking off—as there had been before, as there almost always is in the work of an author advancing in years, especially one, like Dick, who relies on a muse who demands that the work be done swiftly, at high-octane intensity. In terms of his long-term reputation, he died at the right moment, while he was at the peak of his powers and could not subvert his own legend.

And before he could succumb to the temptation, which had always been present but which must have grown greater after the publication of *Valis*, to parlay the muddy revelations of the *Exegesis* into official doctrine and a church, he made his exit. There would surely have been takers, and a clear precedent existed in the success of Scientology. As a spinner of seductive revelations, Dick could generate higher rpm's than Hubbard eight days a week. As to disciples, there was a ready-made audience schooled by two decades of the "drug culture," who only needed to hear from the man's lips that he was, yes indeed, the very Buddha, for things to happen.

To his credit, they didn't. For one thing, he simply wasn't greedy, which seems to be one of the requirements, judging from the examples of Shoko Asahara and L. Ron Hubbard. For another, he was never a good listener. One can't imagine Dick auditing someone else's story; his own was always so much more interesting.

A few days before Good Friday 1997, in a posh suburb of San Diego, Marshall Herff Applewhite, leader of the Heaven's Gate cult, and thirty-eight of his followers committed suicide together in a manner representing the apotheosis of science fiction as a religion in its own right. Since its founding in 1972, when Applewhite met nurse and astrologer Bonnie Lu Nettles at the clinic he'd entered to be cured of his homosexuality, the cult has exemplified a host of themes already touched on in this book. Here is an account of its origins posted on the cult's own Internet Web site:

> In the early 1970s, two members of the Kingdom of Heaven (or what some might call two aliens from space) incarnated into two unsuspecting humans in Houston. The registered nurse was happily married with four children, worked in the nursery of a local hospital, and enjoyed a small astrology practice. The music professor, who had lived with a male friend for some years, was contently involved in cultural and academic activities. . . . They consciously realized that they were sent from space to do a task that had something to do with the Bible.

By 1975 they managed to win national media attention when they enlisted some twenty New Agers who'd wound up in the cul-de-sac of tiny Waldport, Oregon, to follow them to eastern Colorado, where they were to rendezvous with a spaceship that would return them to their home in the stars. Bo and Peep, the two aliens who'd been Applewhite and Nettles, enjoyed a brief celebrity, whose high point was a cover article in the *New York Times Magazine* of February 29, 1976. But when the UFOs didn't keep their promise, fame turned its fickle spotlight elsewhere.

Douglas Curran's book, *In Advance of the Landing: Folk Concepts of Outer Space* (1985), is full of cautionary tales and dismaying photos of bygone UFO celebrities thumbing through their scrapbooks and wearing strange costumes at rinky-dink get-togethers at ghost-town motels. For twenty years that is the life that Bo and Peep (now known as Do and Ti) must have led, acquiring some followers, losing others, and going nowhere. The faith they were selling (at the usual price of "everything a believer has") differed from the gospel according to Whitley Strieber and John Mack of the present time, in that Applewhite's followers were not abducted by aliens but were aliens themselves, their swanlike spiritual essence having been poured into the ugly duckling "vessel" of a human body that had been wandering about the earth ever since, trying to make sense of things.

Like SF fandom, they were a breed apart: unappreciated geniuses whom the world supposed to be mere computer nerds, if that. On the tapes they made just before their communal suicide, one can often hear the tell-tale titter of the SF fan, that smug chortle hinting at some wisdom vouchsafed only to one's fellow fans. Not until the saucer comes to lift up believers can the great joke be shared.

What set apart the Heaven's Gaters from conventional Christian millenarians was their determination not to take no for an answer. Applewhite was not about to spend another twenty five years in the wilderness, unnoticed by the media. In any case, he didn't have another twenty five years. His days were numbered, and Nurse Nettles was already dead. What was taking the mother ship so long?

Finally, like Manson, Jim Jones, and Asahara before him, Applewhite and his whole starship crew had to take the words of Lady Macbeth to heart and screw their courage to the sticking point. Six of them, Applewhite among them, had already demonstrated their capacity for drastic action by having had themselves castrated, a religious practice of great antiquity, though fallen out of fashion in recent centuries. However, there is good authority for the practice in the New Testament, where Christ, asked if it is better not to marry, replies (Matthew 1912): "There are eunuchs who have made themselves eunuchs for the kingdom of heaven's sake. He who is able to accept it, let him accept it." In the third century, Origen, one of the early church's most highly regarded theologians, followed this counsel of perfection and was castrated for heaven's sake. Heaven's Gate, in this respect, is simply part of the gnostic tradition, in which mere matter is of small acount.

Society at large must be grateful for the crucial difference in the path chosen by Heaven's Gate and that followed by Manson, Asahara, and others. The deaths toward which the sect members showed such a philosophical indifference were their own. From the orderliness surrounding their communal suicide, and on the evidence of their videotapes, it would not seem that any of them was acting under coercion. The same can't be said for those who drank the Kool-Aid of Guyana or perished in the flames of Waco. There is even a kind of glory in their deed, a flamboyance of renunciation, as when early Christians marched onto the killing fields of the Colosseum singing hymns.

Gerald Jonas, the long-term SF reviewer for the *New York Times*, wrote a short story in 1970, "The Shaker Revival," which uncannily prefigures the Heaven's Gate phenomenon. Jonas imagines a cult of latter-day rock-'n-roll-inspired ascetics, who leave their homes and jobs, to affirm the four Noes: "No hate, no war, no money, no death." "Try to imagine what real rebellion would be like," one convert explains. Not just another chorus of 'gimme, gimme, gimme . . .' But . . . the Four Noes all rolled up into One Big No." He goes on:

> Our strength is not of this world. You can forget all the tapes and bikes and dances—that's the impure shell that must be sloughed off. If you want to get the real picture, just imagine us—all your precious little gene-machines—standing around in a circle, our heads bowed in prayer, holding our breaths and clicking off one by one. Don't you think that's a beautiful way for your world to end? Not with a bang or a whimper—but with one long breathless Amen?[22]

Often it's easier to respect unconventional behavior in fictional form than its warty, real-life versions. In real life Bo and Peep must have been in some respects conscious charlatans and not wholly taken in by the impostures they practiced on their followers. One of the burdens of leadership (as well as a blessing of being a follower) is such a division of labor. But their faith, just because of its absurdity, embodies the discomforting fact that true religious faith is often a scandal, flying in the face of common sense and inspiring behavior that seems certifiable. As Applewhite had intended, the Heaven's Gate story hit the headlines on Good Friday 1997. It was hard for ministers to avoid the subject when they preached about the Resurrection on Easter. Instead of "Hallelujah!" their message had to be, "Just Say No."

As to the evident absurdities of the Heaven's Gate faith, no less a religious authority than Augustine declared that he believed in Christianity precisely *because* it was absurd. And when two members of the cult who had missed their apotheosis appeared on *Sixty Minutes* on that same Easter evening, they had no more intellectual difficulty shrugging away the silliness of the cult's mythology than Joseph Campbell, and most liberal Protestant theologians, have in shrugging away whatever seems merely mythological in Christianity. What is more pertinent than the literal truth, from a would-be suicide's point of view, is the cost-benefit analysis. Let us assume that the thirty nine believers were not, in the event, taken up by aliens approaching Earth in the shadow of the Hale-Bopp comet. What did they stand to gain—beyond what Hamlet called a consummation devoutly to be wished?

Perhaps just what they were after. At least one of the suicides, Gail Maeder, is reported to have written in her high school yearbook that her ambition in life was "to be rich and famous." If she's lost out on rich, famous has happened—for her and all thirty eight others. Fame is a goal that none of them could have achieved alone, a goal that usually eludes even the very rich and can poison the dainties on their table by its absence. Fame, or false repute, is the motive of most gratuitous lies, and for some of us, it is to die for.

Film rights have already been sold, and there will probably be more than a single movie. It could make a great comedy, after the fashion of *Ed Wood*, or a suspense thriller, or a genuine tragedy.

But it won't be done as science fiction, since that version has already been filmed supremely well as *Close Encounters of the Third Kind.* It was one of the cult's favorite films, and the story of their lives. At the uplifting end of that movie, Richard Dreyfuss defies certain death at the end of the film by entering the supposedly contaminated area around Devil's Tower in order to keep his date with the aliens (who look just like the aliens on the Heaven's Gate Web site). Applewhite and company, having made the same appointment, braved the same fate and have reaped the same reward.

They are immortal.

Chapter 8

REPUBLICANS ON MARS—SF AS MILITARY STRATEGY

Science fiction seldom ventures to speculate about the future of democratic politics. Realities that the news media pay so much attention to have vanished from most fictional futures, where people are governed smoothly and invisibly, as in *Star Trek* imperium, or else their worlds have reverted to a Renaissance Fair era in which simple hierarchies facilitate operatic plotting, as in Gene Worle's *Book of the New Sun* tetralogy or Frank Herbert's *Dune* novels. Readers turn to the op-ed pages for electoral politics or to novelists like Allen Drury, Tom Wolfe, or Anonymous.

But this doesn't mean that SF writers, or their readers, are indifferent to politics. They have their opinions on most issues of the day, opinions that even get some readers fussed, as when Philip K. Dick wrote "The Pre-Persons" (1974), a story in which a mother's right to an abortion extends beyond pregnancy until the "pre-person" reaches age twelve. "In this," Dick wrote, "I incurred the absolute hate of Joanna Russ who wrote me the nastiest letter I've ever received; at one point she said she usually offered to beat up people . . . who expressed opinions such as this."[1] Often, as on college campuses, these intramural controversies seem out of proportion to the opinions expressed because many SF writers hold their peers to a higher standard of liberal orthodoxy.

Of course, not every SF writer is a political liberal. There was a time, in the '40s and early '50s, when liberals certainly predominated among the recognized SF writers. One could scarcely find a brighter shade of pink than that displayed by the Futurians, a loose grouping of East Coast writers that included Isaac Asimov, Damon Knight, Lester del Rey, Judith Merril, Donald Wolheim, and James Blish (most of whom would someday be influential editors as well).

Or take the case of X, a man who one day would be widely regarded as the country's greatest SF writer. In 1938, X ran for the Democratic nomination for the state assembly seat for the Fifty-ninth Assembly District in the state of California. He ran unopposed by another Democrat, but even so he did not make it to the general election, because the incumbent Republican candidate, Charles Lyons, was able to cross-file in the Democratic primary, and, winning under both tickets, was returned to office—thereby terminating X's brief political career.

Lyons won, probably, because of his thirty-one-year-old opponent's close association with EPIC, the quasi-Socialist third party founded by novelist-activist Upton Sinclair. When Sinclair ran for governor in 1934 as a Democrat, conservatives of both parties were so alarmed that an all-out effort was made to stop him. President Roosevelt withheld his endorsement. The movie studios threatened to leave the state. And the future chief justice of the Supreme Court, Earl Warren, declared, "This is no longer a campaign between the Republican Party and the Democratic Party in California. It is a crusade of Americans and Californians against Radicalism and Socialism." X had been the editor of *Upton Sinclair's EPIC News*, a political newsletter with a peak circulation of two million, and one of six men chosen by Sinclair to write a constitution for EPIC in 1935 as it set out to become a nationwide movement. Clearly this young man was no mere fellow traveler and certainly not "the moderate Democrat" he would claim to have been when he once referred to this otherwise deleted section of his curriculum vitae. No, he was the genuine article, a '30s radical leftist, and his name was Robert Heinlein.[2]

Heinlein was scarcely alone in having executed a political about-face between the Depression and the dawn of the McCarthy era. But Heinlein's radicalism, in both eras, was of the homespun variety, the politics of the frontiersman at odds with a federal government it perceived as an occupying power. The economic theories of EPIC stem more from the prairie populism of the 1890s than from Marxist ideology, while his later right-wing preachments were interfused with the libertarianism of a hedonist who doesn't want any of *his* perks infringed. Had he lived so long, he might well have been a Perotista. His one book on practical politics, the rather mild *Take Back Your Government!* posthumously published in 1992, even has a blurb from his most faithful disciple, Jerry Pournelle, declaring, "I would hope that every Perot supporter would read this book prior to the fall campaign."

Even so, the main thrust of Heinlein's SF in the Cold War years was to advocate the perpetuation and growth of the military-industrial complex. He'd graduated from the U.S. Naval Academy in 1928 and would probably have had a career in that service had he not contracted tuberculosis in 1934 and had to retire on a lifetime medical disability. After his failed career in politics, he began publishing SF in 1939, quickly becoming the most popular writer for John Campbell's *Astounding*. At the beginning of World War II, Heinlein took a long sabbatical from SF, returning to work for the navy as a civilian engineer, designing and testing plastics for aircraft and high-altitude pressure suits.

The war years were undoubtedly, for Heinlein as for most of his generation, the defining experience of his mature life. Having braved those deadly terrors, the Cold War never fazed him. Heinlein spoke out against restrictions on nuclear testing in 1956. At a World SF Convention in 1961, he advocated bomb shelters and unregulated gun ownership. He was a hawk in the Vietnam years and fell out with Arthur Clarke on the issue of Reagan's Strategic Defense Initiative. These positions, and others more extreme, may easily be inferred from the SF he wrote in the same period. No hawk could boast sharper talons.

None of his SF novels demonstrates this more clearly than *Starship Troopers* (1959), a paean to the rigors of military life and the ruthless extermination of the enemy, who are, in this case, invading spiderlike alien Bugs. As in World War II, the choice is simple: "Either we spread and wipe out the Bugs or they spread and wipe us out—because both races are tough and smart and want the same real estate."[3]

SF writers have a great advantage in that they can design their own enemies. Heinlein's Bugs, like the aliens in Hollywood's recent version of the same essential story, *Independence Day*, are perfectly vile. Even Nazis, after all, have human children; the children of Bugs are mere worms, or pupae. The climactic scene in *Starship Troopers* is a raid on a Bug nest with the Troopers incinerating millions of evil giant spiders. Soon we'll be able to witness that scene in Technicolor with Dolby sound, for Hollywood FX technology has finally caught up with the written word, and movies can present aliens as convincingly fearful as the insects in our gardens. But the meaning isn't altered. Whether they're aliens or aphids, they must be exterminated.

Aliens are properly the subject of our next chapter. For now let us assume that our human enemies may also be horrible monsters who merit extermination. How best is that to be done? With what weapons? Are there some weapons so deadly that their mere existence will deter an enemy attack?

These questions have a long fictional tradition that begins with a novel that appeared in *Blackwood's Magazine* in 1871—Sir George Tomkyns Chesney's *The Battle of Dorking*—and that has continued to the present day in the work of technothriller writers like Tom Clancy. Chesney, inspired and appalled by the Germans' juggernaut victory over France in the Franco-Prussian War, mapped out, in plausible and horrific detail, a similar debacle for England—if England did not reform its defenses to meet the Prussian threat. Chesney was a career officer in the Royal Engineers, who "followed the tradition of the numerous army and navy officers who had been turning out pamphlets on the problems of national defence ever since the appearance of the steamship had revolutionized the conduct of warfare. But he broke with this tradition by presenting his arguments in fictional form. . . . Soldiers have always had to plan their campaigns on the basis of what they consider the enemy would do in certain circumstances. Add the methods of the realistic narrative to the practice of the military assessment, and the product is a full-scale story of imaginary warfare."[4]

The Battle of Dorking produced a flood of imitators over the next forty years. That such books were aimed not simply at producing royalties for their authors and publishers but at persuading the taxpaying public that their nation's survival depended on increased expenditure was clear from the number of admirals, generals, and politicians who turned their hands to the new genre. Fiction became an instrument of public policy:

> By 1906 all the varied effects of the new mass dailies, universal literacy, the growth of armaments, and the ending of British isolation came together in the episode of the *Invasion of 1910*. With the agreement and active encouragement of Lord Northcliffe, Field-Marshal Lord Roberts worked in close association with the popular journalist, William Le Queux, in preparing a story about a German invasion of Britain in 1910 for serialization in the *Daily Mail*. Lord Roberts saw it as an opportunity to spread his conviction that Britain must be better prepared for a modern war.[5]

Americans were not at first notable among the pioneers of the new genre, since the country was still busy subduing its own continent and had two oceans to serve it as a system of defense. There was one account, *The Stricken Nation*, by Henry Grattan Donnelly (1890), of a British invasion that destroys most of the cities on the eastern seaboard, but it was another Donnelly who, one year earlier, had written the first distinctively American account of a future war—in this case a civil war, not of North against South but of rich against poor. The book was *Caesar's Column: A Story of the Twentieth Century*, and its author was, like Heinlein, another failed populist politician from the Midwest *and* another Donnelly.

We've encountered Ignatius Donnelly already, as the author of *Atlantis* and *Ragnarok*, those two seminal and enduring classics of pseudoscience. *Caesar's Column*, too, was a best-seller in its time, but it has not managed to stay in print, since prophecies of imminent doom lose their appeal once they have been overtaken by events. It is all the more astonishing, then, that Donnelly prefigures in so many ways the work of Heinlein and his followers. Since it is very unlikely that any of them have read *Caesar's Column*, the resemblance must be ascribed to their common origins in what Richard Hofstadter has termed "the paranoid style in American politics,"[6] a tradition that sees sinister machinations in the operation of compound interest and imagines the night skies abuzz with black helicopters. The core of this tradition is its ambivalent feelings toward the power of the federal government, sometimes celebrating its military might and cheering on the westward course of empire, at other times resenting that power when it is exercised to tax their incomes or regulate their private arsenals.

Caesar's Column describes a dystopian America ruled by a corrupt oligarchy and served by a brutalized working class, whose sorry fate inspires Donnelly to passages of high and purple oratory:

> Toil, toil, toil, from early morn until late at night; then home they swarm; tumble into their wretched bed; snatch a few hours of disturbed sleep, battling with vermin, in a polluted atmosphere; and then up again and to work; and so on, and on, in endless, mirthless, hopeless round; until in a few years, consumed with disease, mere rotten masses of painful wretchedness, they die, and are wheeled off to the great furnaces, and their bodies are eaten up by the flames, even as their lives have been eaten up society.[7]

In both its abhorrence of the ruling class and his sympathy for exploited workers, Donnelly's novel bears a strong and surely not accidental resemblance to Jack London's *The Iron Heel* (1907). But it differs from its successor in some telling ways—most strikingly by that paradoxical anti-Semitism that supposes Jews to be in control of both Wall Street and international bolshevism. We learn that when Donnelly's archvillain, Prince Cabano, "comes to sign his name to a legal document, [he] signs it as Jacob Isaacs." And further, that "the aristocracy of the world is now almost altogether of Hebrew origin." But the second in command of the revolutionary Brotherhood of Destruction, and the "brains of the organization," is a Russian Jew, "driven out of his synagogue in Russia . . . for some crimes he had committed . . . a man of great ability, power and cunning." When the revolution has brought about its own Reign of Terror and the destruction of New York together with a quarter-million of its citizens, this Jewish Robespierre flees to Judea, where "he proposes to make himself king of Jerusalem, and, with his vast wealth, re-establish the glories of Solomon, and revive the ancient splendors of the Jewish race, in the midst of the ruins of the world."

The most notable SF feature of *Caesar's Column* is its anticipation of the horrors that would come to characterize twentieth-century aerial warfare, from Guernica to London and Dresden and on to Vietnam:

> The Oligarchy have a large force of several thousands of these [dirigible airships], sheathed with that light but strong metal, aluminum; in popular speech they are known as *The Demons*. Sailing over a hostile force, they drop into its midst great bombs, loaded with the most deadly explosives, mixed with bullets; and, where one of these strikes the ground, it looks like the crater of an extinct volcano; while leveled rows of dead are strewed in every direction around it. But this is not all. Some years since a French chemist discovered a dreadful preparation, a subtle poison, which, falling upon the ground, being heavier than the air and yet expansive, rolls . . . steadily over the earth in all directions, bringing sudden death to those that breath it. . . . It is upon this that [the Oligarchs] principally rely for defense from the uprisings of the oppressed people.[8]

In this vision of the government's employing weapons of mass destruction against its own people, Donnelly prefigured a number of later dystopian writers from Jack London to George Orwell. Usually these writers were of a left-liberal bent who feared their own rulers at least as much as the enemy. In the aftermath of both world wars and of genocides in Turkey, the USSR, Germany, China, Cambodia, and throughout the Third World, the nightmarish possibility that any government might simply exterminate whole classes of its citizens came to be the common property of both right and left. The nuclear logic of mutual assured destruction dominated the military planning of Cold War strategists, and from the point of view of those citizens situated at ground zero it mattered little *whose* bombs ensured their destruction. They were the hostages of their government's policy. Even those "loyal" enough to affect the bravado of slogans like "Better Dead Than Red" experienced the dread of living in the shadow of the atomic sword of Damocles. To the degree that they could identify with those who wielded that sword, they were hawkish, calling for the use of atomic weapons in every international crisis.

Heinlein was such a hawk, convinced that only a Pax Americana backed up by nuclear superiority could secure world peace, as he proposed in his

1941 story, "Solution Unsatisfactory." In *Starship Troopers* he gives us glimpses of an Earth ravaged by the Bugs' invasion but still indomitable, a scenario that appears as a copy in the movie *Independence Day,* in which the American president is allowed to speak the words tabooed to all previous movie presidents: "Let's nuke the bastards!"

In the mid-'70s, as Heinlein lapsed into aesthetic solipsism with novels like *Time Enough for Love* and *The Number of the Beast,* books in which the Heinlein persona multiplies itself like a computer virus until the very universe becomes just a figment of his imagination, his mantle as SF's premier cold warrior fell upon the ample shoulders of Jerry Pournelle.

Pournelle, like Heinlein, began his intellectual career as a radical leftist, a member, indeed, of the Communist party. He accounted for this aberration in a 1981 interview with Charles Platt:

> That [his party membership] was a long time ago. After I got out of the Korean war, and came back and was an undergraduate, I fell into the hands of those who kept telling us that Marxism was within the Western tradition, and so forth. I was also a victim of the snigger-theory of philosophy, which is that if you admire anyone other than a leftist then you're barely tolerated in the university department, and they laugh at you. I had been through a pretty miserable war, the communists promised to do something, and it didn't look to me as if anyone else was going to do anything. [He shrugs.] Misplaced idealism.[9]

Pournelle next misplaced his idealism in the aerospace industry, where

> I had a very senior position for someone my age, in North American Aviation, which at that time was the outfit that was building Apollo. I was a Space Scientist; my position was to find things within the company that I thought I could contribute to, and go work on them. The last professional assignment I had was to work on the experimental design for Apollo 21. But there wasn't going to be any Apollo 21, it became fairly obvious, and at the same time the management said, "We've got to trim the number of people who are senior scientists."[10]

Having been trimmed, Pournelle commenced his career as a freelance writer, publishing his first SF story, "Peace with Honor," in 1971. He worked as a director of research for the city of Los Angeles and wrote speeches for the mayor. He became the secretary of the L5 Society, a group advocating sending colonists into outer space to live in great orbital hives. He was a cofounder of the Citizen's Advisory Council on National Space Policy, an early advocate of Reagan's Strategic Defense Initiative (SDI). He is a regular columnist in PC magazines (personal computers, that is, not the other PC)—and a contributing editor to *Soldier of Fortune*, the magazine for would-be mercenary soldiers.

Pournelle's first success in SF came from his CoDominium series, featuring John Christian Falkenburg, who commands a star-fleet of latter-day Renaissance mercenaries. Falkenburg and his legion are compellingly imagined—amoral (excepting among themselves), competent professionals. In his first adventure at novel length, *The Mercenary* (1977), Falkenburg brings peace to an entire planet by assembling all the local welfare-supported riffraff into a sports stadium and blowing them up. One senses that the author would not be averse to exercising a similar cleansing initiative on his home planet.

Ben Bova, the SF editor who published Pournelle's first Falkenburg story and its successors, had been, like Pournelle, an aerospace employee before he began to write and edit SF. While Pournelle is a would-be mercenary, Bova is a would-be astronaut, who has championed the space program with an ardor and intelligence second to none. His nonfiction book, *The High Road* (1981), argues that a civilian space program should be a top national priority, but he also concedes that NASA and the military are joined at the hip. "Be aware," he counsels, "that there will always be a military presence in space, as long as nation can lift sword against nation. But if we can begin to ease worldwide tensions by bringing the wealth of energy and raw materials from space to the peoples of the world, then we may be able to avert the military buildup that could lead to Armageddon."[11] This same argument would be the basis for Bova's advocacy of Reagan's SDI, *Assured Survival: Putting the Star Defense Wars in Perspective* (1984).

From the time that the German space enthusiast, Wernher von Braun, was recruited to develop the V-1 and V-2 rockets for the Nazis, there has been a correspondence approaching identity between the exploration of space and the development of superweapons. By the early '80s, the arms race had come to a seeming dead end that guaranteed an intolerably precarious standoff based on mutual assured destruction. At that point, science fiction came to the rescue with a new formula, mutual assured survival. That was, in fact, the title of a tract written by Jerry Pournelle and another SF writer, Dean Ing, the back cover of which boasted a letter of recommendation from President Reagan. Reagan's letter offers little more than "Thank you, and God bless you," but implicitly it endorses the authors' recommendations:

> We believe it imperative that we first address four candidate systems which provide a significant military capability, i.e., to deny assurance of first-strike success by any aggressor by 1990.
>
> - Multiple satellite using kinetic energy kill.
> - Ground-based lasers and mirrors in space.
> - Space-based lasers.
> - Nuclear explosive-driven beam technologies collectively known as third generation systems.
> - Ground-based point defense system.
>
> We also urge greatly accelerated research on the many other candidate systems, including particle-beam weapons, which offer promise on the longer term.[12]

Pournelle and Ing's book did not raise them to the status of media pundits, having been published as a paperback original by a publisher outside the mainstream, Baen Books, specializing in SF of the Heinlein/Pournelle persuasion. Indeed, the text of *Mutual Assured Survival* reads like a set of specs designed to provide technological background for the latter-day space operas of Pournelle, Ing, and such other Baen authors as David Drake (*Hammer's Slammers* [1979], and its many sequels) and William R. Forstchen, author of *Star Voyager Academy* (1994) and Newt Gingrich's "coauthor" for the novel *1945*.[13]

To the degree that SDI was successfully ballyhooed to American voters, the job was accomplished not by the written word, nor yet by the President's earnest advocacy, but by the computer animators who designed the film clips that were broadcast on television news programs week after week beginning in 1983. The first computer-animated feature film, Disney's *Tron*, had appeared only the year before. High-quality computer graphics did not then register as "cartoonish" in the Mickey Mouse sense but rather as a new "virtual" reality. They had the sheen of the cyberspace future and were employed in those SDI publicity clips to sell voters a new product, one that promised to trump the enemy's intercontinental ballistic missiles and, less doubtfully, to justify new levels of Pentagon and aerospace expenditure. None of the systems so awesomely animated actually existed in any form other than as science fiction or arcade games.

It's doubtful that we will ever know, now that Reagan has entered his own private twilight zone, the degree to which SDI was a serious proposition or an inspired hoax. As a serious proposition, it has never been a great success, having met with derision from scientists with no vested interest in its development. The first tests of prototypes were shown to have been gimmicked, but even the gimmicked results were accounted unpromising. As a hoax, however, SDI may have played a significant part in ending the Cold War. It wasn't just the electorate those computer graphics persuaded: the Soviet government also took Reagan at his word and became convinced that SDI would initiate another round of spending to maintain the balance of terror. In Reagan's second term, the Russians gradually loosened their grip on Poland and the rest of Central Europe. Like a tired marathoner who has reached his limit and gears down to a jog that represents an honorable surrender, the Soviet Union exited the arms race.

Reagan's big bluff may have been the one lasting accomplishment of his administration, but it was scarcely an unalloyed triumph. As Garry Wills has observed: "It is claimed that Reagan brought down the Soviet Union by forcing it to spend so much on defense that it collapsed. Perhaps. But in the process he made *America* spend so much on its military that the budget deficit nearly tripled and the trade deficit more than quadrupled. . . . Even when the cold war was over, few in America seemed to be celebrating. It is not given to many Presidents to spend two world empires toward decline."[14]

It's doubtful that anyone in the Reagan administration would have consciously initiated a program of national self-sacrifice on the scale Wills suggests. More likely, proponents of SDI, from the President down to the mercenaries in the field like Pournelle, were aiming at something simpler: the continuing or enhanced prosperity of their coworkers in the military-industrial complex, something that the proposed spending on SDI was guaranteed to accomplish.

SDI was allowed to fade from the forefront of Reagan's agenda after the *Challenger* disaster on January 28, 1986. High-tech solutions suddenly seemed high-risk. If the shuttle's six astronauts and teacher Christa McAuliffe could die as the result of a defective O-ring, perhaps it might not be wise to let the fate of cities and continents depend on orbiting Rube Goldberg mechanisms, no matter how well attested their safety and efficacy. Was HAL of *2001* to be put in charge of the entire planet's nuclear fate?

The ninety-five-member astronaut corps, including the chief astronaut, John W. Young, Jr., was shown to have their own grave doubts about NASA's concern for safety. Their doubts were routinely dismissed. "One veteran astronaut," the *New York Times* reported on April 3, 1986, "cited the case of an astronaut who, when he voiced safety concerns, was humiliated by his superiors in front of other people as being afraid to fly. . . . Another senior astronaut . . . said that he, too, felt that astronauts risked retribution by voicing safety concerns. Crew selection, he said, was done not by the astronaut office by by higher NASA officials." Later in April, reporting on an investigation by the General Accounting Office, the *Times* ran a series on NASA's wasteful spending *and* its indifference to safety. Further shuttle flights were postponed for the duration of NASA's embarrassment, and the entire rationale for manned missions into space was called into question by prominent scientists.

At this juncture, when NASA found itself on the defensive for the first time in its existence, the *New York Times* published a full-page "Letter to the American People" that took up the agency's fallen and bloodied banner and urged that the most fitting memorial for the seven dead would be

> the restoration and enhancement of the shuttle fleet and resumption of a full launch schedule.
>
> For the seven.
>
> In keeping with their spirit of dedication to space exploration and with the deepest respect for their memory, we, the undersigned are asking you to join us in urging the President and the Congress to build

a new shuttle orbiter on the work of these seven courageous men and women.

AS LONG AS THEIR DREAM LIVES ON
THE SEVEN LIVE ON IN THE DREAM.[15]

The letter was signed by eighty-eight "underwriters" and some two hundred others listed in small typeface, each of whom contributed $350 toward the ad's $36,000 cost. Prominent among the underwriters were Isaac Asimov, Robert Heinlein, and some few other brand-name SF and fantasy writers, but the extent to which the list was composed of SF professionals—writers, editors, agents—would be evident only to other members of the club: by my own census, at least fifty-four of the eighty-eight underwriters, with the Heinlein/Pournelle contingent present in full strength.

NASA, with SF as its unofficial PR agency, had for years counted on manned space flight to win public support for its budgetary requirements. The astronaut corps humanized the space program, an endeavor that otherwise is the epitome of science at its most inhumane: not cruel but rather loftily indifferent to human concerns. Its greatest boon has been the satisfaction of our intellectual curiosity about a solar system that all NASA's probes have shown to be lifeless. The recent discovery of what are said to be fossil evidences of bacteria on Mars some millions of years ago was met by NASA loyalists with cries of jubilation. Life on Mars! Reason to mount an expedition to bring back *conclusive* fossils.

In H. G. Wells's *The War of the Worlds*, the invading Martians were destroyed by Terran bacteria against which they had no immune defense; today the only Martians for which we have a shred of evidence are germs, now probably extinct. Any "sense of wonder" to be derived from this is of very dark hue. No doubt there would be further dark insights to be gained from more close-up photographs of the planets and asteroids that are neighbors of Earth, but there is little reason to suppose further exploration will yield a theological bonanza in the form of a true alien Other.

The space program has proved altogether a hard sell. If even those who once thrilled to the *Apollo* landings are of two minds about pumping more money into NASA, there are others for whom NASA is simply a rival species in a Malthusian battle for government funds. "I wonder how many people out there feel as I do?" a correspondent to my local newspaper writes in a letter to the editor. "I refer to the recent decision to explore Mars and Venus for life. How many millions of dollars will be spent on this? In the meantime people right here on Planet Earth are starving and homeless. Our elderly people are just scraping by—why not keep the money here—on our planet and put it to good use?"[16]

A more revealing expression of the hostility many feel toward the space program can be heard in Gil Scott-Heron's song, "Whitey's on the Moon," or in this account by a black academic who teaches at the University of Pennsylvania. She tells of how, while visiting in-laws in Georgia, she'd witnessed the TV coverage of the destruction of the *Challenger*:

> I like most Americans, had a special concern for the fate of one crew member—the appealing young white woman schoolteacher, Christa McAuliffe, who had won her chance to travel in space in a national competition for teachers. I said that I was sorry about her death. My remark was met only by blank stares—in fact, all of the commentary about the accident was received with total disinterest by my in-laws, except for one bit of information. When the name and photo of the lone black crew member, Ronald McNair, were flashed on the screen, they leaned forward with a single coiled motion, galvanized into intense attention. This one alone was ours, their posture reminded and rebuked me; this one was our concern, this man alone deserved our mourning.
>
> As we watched *Challenger*'s destruction, some of us who had muttered, "God will not be mocked," at the hubris of its name were sad and respectful, simply to mourn Ron McNair. If he had not been aboard, perhaps a jubilant chord would have been struck that day; perhaps a ballad like "Titanic" (the one about Shine or that other one that used to be off-timed by blacks with the shout of "Hallelujah!" following "It was sad when the great ship went down") would have resulted.[17]

One must respect the candor of this statement, however one may deplore (or seek to mitigate) its overt racism. It is, unfortunately, an accurate refletion of the degree to which the space program has become, as much as abortion, a partisan issue. The right is pro, the left con. And under the Clinton administration, the balance has tilted con-wards. In May 1995 NASA announced that it would be making deep cuts in its workforce as its budget shrinks "in a new era of austerity." Twenty to 30 percent of civil service and contractor positions would be eliminated, just as the underwriters of the letter to the *Times*, a decade earlier, had foreboded. "After the cuts," the *Times* reported, "NASA would be the size it was in 1961, before the nation began the *Apollo* program."[18]

SDI had already become the ghost of the chimera it had been, thanks to the self-destruction of the Soviet threat that it had been designed to confront. Without a nuclear enemy of sufficient magnitude, the immense and unproven apparatus of SDI was not a salable product, and NASA was, once again, the last, best hope of the aerospace industry. At this juncture an unlikely would-be savior stepped into the breach.

Newt Gingrich had long been one of NASA's most faithful boosters. He had also been, for some time, employing writers from the Pournelle/Baen Books stable of space opera commandos as ghosts and cowriters. Gingrich's first significant SF collaborator was the immensely prolific David Drake, author of *Hammer's Slammers* and its many sequels concerning mercenary soldiers of future interstellar wars. Their book, *Window of Opportunity: A Blueprint of the Future*, was published, to little attention, in 1984 by Baen Books and prefaced by Jerry Pournelle, who salutes Gingrich's (and Drake's and the third coauthor, Marianne Gingrich's) work as "a remarkable book, almost unique in that, without the slightest compromise with the principles that make this nation great, Gingrich presents a detailed blueprint, a practical program that not only proves that we can all get rich, but shows us how."[19]

Adventures in space turn out to be a major component of what we are to hope for. One can't help but sense the influence of Gingrich's SF connections (Drake/Pournelle/Baen), especially in passages like this: "Congressman Bob Walker of Pennsylvania [now chairman of the House Committee on Science, Space, and Technology] has been exploring the possible benefits of weightlessness to people currently restricted to wheelchairs. In speeches to handicapped Americans, he makes the point that in a zero-gravity environment, a paraplegic can float as easily as anyone else. Walker reports that wheelchair-bound adults begin asking questions in an enthusiastic tone when exposed to the possibility of floating free, released from their wheelchairs. Several have volunteered to be the first pioneers."

This "Arise and float!" is evangelism with a canny subtext, not unfamiliar to SF professionals. Outer space is envisioned as that New Frontier where the indignities of ordinary life—onerous no-future jobs and low status—are to be remedied, as they were by an earlier expansion into the American West. Space is like Texas, only larger. In the twenty-first century, Gingrich (or his SF ghost) declares, a third-generation space shuttle "*will* be the DC-3 of space. From that point on, people will flow out to the Hiltons and Marriotts of the solar system, and mankind will have permanently broken free of the planet."

That was 1984, when the name of Gingrich did not bulk large. Since then he has become a legend (though already the legend has faded greatly), partly for spearheading the 1994 Contract for America campaign that gave Republicans their first House majority in forty years, but also for using his political celebrity to make record-setting book deals. For his 1995 manifesto, *To Renew America,* HarperCollins offered a $4.5-million advance, but the contract had to be withdrawn when Democrats in Congress went ballistic with righteous indignation.

In *To Renew America* Gingrich continued to beat the drum for the space program and managed to put in a good word for an old friend:

> Jerry Pournelle, an aerospace engineer and science fiction writer, notes that going into orbit takes about the same amount of energy as traveling from Los Angeles to Sydney, Australia. . . .
>
> I believe that space tourism will be a common fact of life during the adulthood of children born this year, that honeymoons in space will be the vogue by 2020. Imagine weightlessness and its effects and you will understand some of the attractions.[20]

But this doesn't mean that Gingrich is a steadfast supporter of NASA. It was an article of faith with Heinlein that space should be strip-mined by private enterprise, as per the title of his first collection of stories, *The Man Who Sold the Moon* (1950). Government should not be trusted with such a vital mission, and Gingrich explains why in terms of strict Heinleinean orthodoxy:

> One of the major reasons the spirit of adventure has gone out of space exploration is that we have allowed bureaucracies to dominate too many of our scientific adventures. Look at the difference between the movies *The Right Stuff* and *Star Wars* and you will see what I mean. Bureaucracies are designed to minimize risk and to create orderly systematic procedures. In a way they tend to be inherently dull. They also tend to be slow, cumbersome, and expensive. . . .
>
> I recently asked an aerospace executive what would happen if we got the government out of the business of designing space shuttles and space stations. She replied that the cost would drop by 40 percent and the amount of time would be cut in half.[21]

One wonders if the seven who died in the *Challenger* explosion would find this entirely persuasive.

For the SF faithful, though, the important thing is to have a fantasy that is sufficiently persuasive and brightly colored. The future is a theme park that allows us to enter our favorite SF movies and have cheerful adventures. And that's no metaphor:

> Our generation is still seeking its Jules Verne or H. G. Wells to dazzle our imaginations with hope and optimism. Instead we have writers like Michael Crichton, whose work is just standard alarmist environmentalism in which humans are forever messing up nature, a Frankenstein story in which curious scientists are the villains.
>
> Why not aspire to build a real Jurassic Park? (It may not be at all impossible, you know.) Wouldn't that be one of the most spectacular accomplishments of human history?[22]

Gingrich's enthusiasm for SF as the best vehicle for promoting the prosperity of SDI and the space program has led him to his own attempts in the genre. For some time he has been working on a collaborative future-war novel, with that veteran of the genre, Jerry Pournelle, to be published by Baen Books. Its tentative title is *The Faction*, and its premise, in the tried-and-true vein of *The Battle of Dorking*, is a sneak attack by the Enemy (in this case, Japan and its minions) against the United States, using SDI-style superweapons. At last report, 120,000 words of the book had been written, but according to publisher Jim Baen, the book remains unfinished and "in need of drastic revision because of some kinks in the plot." Given its return-of-the-Yellow-Peril plotline, I'd lay odds that it will remain unpublished so long as Gingrich is Speaker of the House.

Gingrich was happier in his choice of enemies for the one SF novel that he has published. That novel, *1945*, was published by Baen Books and coauthored by another SF writer from the Baen stable, William R. Forstchen. *1945* posits an alternate past in which Hitler did not declare war on the United States in the aftermath of Pearl Harbor, and so America achieved an early success in the Pacific and Hitler conquered all Europe. In 1945, the two world powers are in a race against time to develop and deploy the atom bomb. The novel ends with a long, impassioned plea for unrestricted secret budgets for the aerospace industry, a message much the same as that of *The Faction*, though more palatable when one's enemies are as unimpeachably villainous as Hitler and the Third Reich. Even so, this may be another unfinished story, since the sales of *1945* were spectacularly dismal, and the sequel promised by the "To Be Continued" on the last page might prove to be an unredeemed pledge. Few tasks can be so joyless as writing Books 2 and 3 of a trilogy after Book 1 has bombed.

1945's flop should not be taken as a sign of the waning of either the Pournelle/Baen axis in SF or of Gingrich in national politics. It was simply a marketing miscalculation. In 1994, in a rare moment of self-deprecation, Gingrich said of the novel in progress that it was "mostly for the Arnold Schwarzenegger group." With all due respect for Schwarzenegger's charismatic screen presence, I take that to mean "dopes." If the author has so cynical an opinion of a book to which he has lent his name, the reading public may be excused for sharing it.

As for the SF reading public, which is now large enough that at times three or four slots on the best-seller lists are occupied by SF titles, the book presented other difficulties. It sounds retro. The action takes place, after all, in the black-and-white past, and, as in *The Man in the High Castle* and several other "alternate histories" in both SF and the mainstream, the Nazis have won and World War II has to be replayed. The tone of the book is retro as well, having had the spicier elements of its first draft eliminated in deference to Gingrich's political persona as a champion of family values. All that's left for "the Arnold Schwarzenegger group" to enjoy are the weapons, and they too are retro, despite their "dark, compelling attraction . . . like being aroused by a woman one despised."

Even among conservatives, time marches on, and the warrior fantasies of Pournelle and Forstchen have been shouldered aside by work that is more viscerally violent, sexually explicit, and politically dismaying. Among the younger writers, Gingrich's collaborator, William Forstchen (born in 1950), is probably the most old guard, but even he writes from the paranoid perspective of the militia movement. His 1994 *Star Voyager Academy* has a cover designed to resemble a recruiting poster for NASA, complete with the Stars and Stripes. The plot, cloned from Heinlein's *The Moon Is a Harsh Mistress*, concerns the beginning of a new Revolutionary War as the stalwart space colonists of Mars take on the oppressive forces of the U.N. The overall tone is mild, but the moral of the story is distinctly post-Vietnam rather than post–World War II: the Government is your enemy. No taxation without a military coup d'état. Beware of black helicopters.

This has been the orthodox viewpoint for many years along the Baen/Pournelle axis, and is now common among extreme right-wingers, especially those of the militia movement. To cater to such readers, a new subgenre of military SF has developed that celebrates nuclear war and its anarchic aftermath as the crucible in which future Rambos may be formed. These books invariably appear as open-ended series: Jerry Ahern's Survivalist series (twenty-four volumes between 1981 and 1992; from Zebra), the Traveler series by D. B. Drumm[23] (thirteen volumes between 1984 and 1987; from Dell), the Ashes series by William Johnstone (thirteen volumes between 1983 and 1991, from Zebra), and, without exaggeration, many, many more. What all these series have in common, besides their postholocaust settings, are salacious violence and devout self-righteousness. Here's how the critic John Clute has characterized Johnstone's Ashes series: "The premise of the first volume is, perhaps, surprisingly frank: shocked by the imposition of gun control, a group of patriotic US citizens bring about the nuclear holocaust in the expectation that a better world will, phoenix-like, be born. The remaining volumes of the sequence attempt to demonstrate how right they were."[24] (Any resemblance to the Oklahoma City bomber Timothy McVeigh may not be entirely coincidental, but of McVeigh and his specific SF connection, more below.)

For confirmed readers of such series (and there are, alas, legions), the literary efforts of Gingrich/Forstchen and even Pournelle must seem like weak tea, if not tap water. But once again, these survivalist fantasies have a clear precedent in the work of that grand master of the genre, Robert Heinlein, who set forth his theory of the social and psychological benefits of nuclear war in *Farnham's Freehold* way back in 1964. In that book, it's zero hour, and Hugh Farnham, our hero, has been foresighted enough to build a fallout shelter. As he and his family settle in after the blast, Hugh philosophizes about the long-term benefits of their situation:

> "Barbara [he says to his adoring spouse and straight man], I'm not as sad over what has happened as you are. It might be good for us. I don't mean us six. I mean our country."
>
> She looked startled. "How?"
>
> "Well—It's hard to take the long view when you are crouching in a shelter and wondering how long you can hold out. But—Barbara, I've worried for years about our country. It seems to me that we have been breeding slaves—and I believe in freedom. This war may have turned the tide. This may be the first war in history which kills the stupid rather than the bright and able—where it makes any distinction."
>
> "How do you figure that, Hugh?"
>
> "Well, wars have always been hardest on the best young men. This time the boys in service are as safe or safer than civilians. And of civilians those who used their heads and made preparations stand a far better chance. Not in every case, but on the average, and that will improve the breed. When it's over, things will be tough, and that will improve the breed still more. For years the surest way of surviving has been to be utterly worthless and breed a lot of worthless kids. All that will change."
>
> She nodded thoughtfully. "That's standard genetics. But it seems cruel."[25]

Not all SF of a military stamp is of the School of Heinlein with its insistence on the evolutionary superiority and inevitability of a warrior ethos. Gordon Dickson, Poul Anderson, and Keith Laumer have all written novels that deal with the war in a thoughtful manner, while remaining essentially within the boundaries of genre SF and offering the high-tech thrills of space opera. Their best work has the epic scale and ethical weight of first-rate popular historical fiction by such novelists as Gore Vidal, Cecelia Holland, and Gary Jennings. Significantly, all three of those authors have themselves written SF (Holland's *Floating Worlds* [1976] is an epic-length space opera of unusual merit), while both Dickson and Anderson have written a good amount of historical fiction.[26]

For much of this century, warfare was considered the greatest theme a novelist might aspire to. In part, this was due to its prominence in the epics of classic literature and to the cachet of Tolstoi's *War and Peace*, which many (including myself) believe to be the greatest novel ever written; in part, to those who achieved early literary fame in our time with novels reflecting their own experience of war (Remarque, Hemingway, Mailer, Jones, Wouk); and, setting all literary considerations aside, because war has been a constant and dire presence in the daily lives and imaginations of most residents of the century.

SF has produced few novels about war that rise to the level of the best realistic depictions of those who have endured the experience and lived to write about it. The conventions of space opera derive from the melodramatic stereotypes of pulp fiction and a frame of mind for which the kindest adjective is *puerile*. The great-grand-daddy of the composers of space operas, E. E. ("Doc") Smith, set a jingoist tone that can still be heard reverberating in the work of Heinlein, Pournelle et al. In *Skylark of Valeron* (1935) his hero perorates in this wise:

> "I'll bet my shirt that the chlorins are wiping out the civilization of that planet—probably folks more or less like us. What d'you say, folks—do we declare ourselves in on this, or not?"
>
> "I'll tell the cockeyed world . . . I believe that we should. . . . By all means . . ." came from Dorothy, Margaret, and Crane.
>
> "I knew you'd back me up. Humanity *uber alles—homo sapiens* against all the vermin of the universe!"[27]

This is to Tolstoi as bubble gum is to bouillabaisse, but it is an article of faith among SF purists, like Heinlein's idolaters, Alexei and Cory Panshin, that E. E. Smith is somehow superior to "mundane" writers like Tolstoi, precisely because he expresses such lofty sentiments.

For all that, the genre *has* produced novels about war that are as vivid, as resonant, and as worthy of acclaim as any other written in this century—among them: George Orwell's *1984*, Kurt Vonnegut's *Slaughterhouse Five*, Walter Miller's *A Canticle for Leibowitz*, and Joe Haldeman's *The Forever War*. The first three have already attained that degree of celebrity that their admirers can protest that they are too good to be accounted SF.

In the fullness of time, I'd prophecy, the same will be said for *The Forever War*. Haldeman served from 1967 to 1969 as a combat engineer in Vietnam, where he earned that least enviable of honors, a Purple Heart. *The Forever War* was published, in novella-size portions, between 1972 and 1974 in *Analog Science Fiction*, then under the editorship of Ben Bova (editorially, his proudest hour), and went on to win, in book form, both the Hugo and Nebula awards. It deserved a Pulitzer, for it is to the Vietnam War what *Catch-22* was to World War II, the definitive, bleakly comic satire. It is also a refutation of all the guff about guts and glory purveyed by the School of Heinlein. As with *Catch-22*, or for that matter *War and Peace*, it defies synopsis or paraphrase. Its basic premise is that future interstellar warriors are propelled to their theaters of operation by "collapsar jumps"—for them an overnight jaunt but, objectively, spanning decades or longer. Their alienation from the command structure, and the cause they presumably serve, is all but total, and any relation to the experience of combat soldiers in Vietnam is ever more tellingly on target as the novel deepens.

The problem with all masterpieces is that they are by their nature unrepeatable. Unlike the various survivalist series and the Soldier of Fortune adventures of Pournelle, Drake, and Co., *The Forever War* said what it had to say once, and unforgettably. It is but a single book among entire racks of paperbacks. But if you believe the legend of Rambo, one staunch warrior who is pure of heart can defeat whole squadrons.

Chapter 9

THE THIRD WORLD AND OTHER ALIEN NATIONS

If the rocket ship is the basic icon of science fiction, the genre's primal scene is first contact with an alien, whose existence is living proof that we are not alone. If God can't be coerced into breaking his silence, at least he can send emissaries. And if the written record is any indication, he has been doing so long before UFOs began messing with Whitley Strieber.

The Swedish mystic and mineralogist Emanuel Swedenborg wrote one of the first eyewitness accounts of a close encounter of the third kind. In his treatise of 1758, *De telluribus* [Concerning Other Worlds], Swedenborg explained how he became acquainted with the inhabitants of Mercury, Venus, Mars, Jupiter, and Saturn:

> Whereas I had a desire to know whether other Earths exist, and of what sort they are, and what is the nature and quality of their inhabitants; therefore it had been granted me of the Lord to discourse and converse with spirits and angels . . . from other Earths; with some for a day, with some for a week, and with some for months.[1]

Mercury was inhabited by simple farmers of good character and orthodox Christianity. There were more good Christians on Venus, but also a rougher breed of alien—giants so tall that "the men of our Earth reach only to their navels. Also . . . they are stupid, making no inquiries concerning heaven or eternal life, but immersed solely in earthly cares, and the care of their cattle."

The trance-voyaging philosopher gave a very circumstantial account of the Martians he met. Though beardless, "the lower region of his face was black," while "the upper part of the face was yellowish, like the faces of the inhabitants of our Earth who are not perfectly fair." The Martians subsist on a vegetarian diet of fruit and beans, and their clothes are made of tree bark woven and gummed together. The men of Jupiter dressed similarly and reposed on beds of leaves in conical tents resembling tepees. Saturnians, like Jovians, were mostly God-fearing Christian aliens, except for one idolatrous minority that worshipped the planet's rings.

As a recent commentator on *De telluribus* observes, Swedenborg, writing more than a century before space flight was imagined, "did not visualize the alien as a traveler geared and helmeted for space flight [but rather] . . . as an inhabitant of remote space, modeled upon those who were familiar to him."[2] The "remote space," in this case, is the New World, as known to Swedenborg from the reports and apocryphal tales of the first explorers. The giants of Venus are modeled on the Patagonians of South America described by Magellan's fanciful chronicler, Antonio Pigafetta, who witnessed "a naked man of giant stature on the shore of the port, dancing, singing and throwing sand on his head. . . . He was so tall we reached only to his waist." Later reports held the Patagonians to be crudely developed cattle herders of bestial stupidity. The characteristics ascribed to Martians, Jovians, and Saturnians—their beardlessness, tawny complexions, simple diet, and habitations—are similarly derived from what had been reported of (or imputed to) North American Indians.

These are very humble beginnings for what was to become SF's most versatile metaphor, its signature trope, the Alien, but they are no less significant for their simplicity. For what can any of us suppose about what is unknown to us, except by reference to the nearest thing it might resemble? If the planets are worlds like ours, then will not their inhabitants be people like us? So long as creation was accounted the act of a God who wanted to mold some beings in his own likeness, it stood to reason that any extraterrestrial acts of creation would vary no more from the basic pattern than do the varying races of humankind. This assumption held steady to the end of the nineteenth century, until Darwin's new account of human origins made it seem likely that if natural selection rather than God was in charge of life on other planets, nature might be red in tooth and claw throughout the universe. There might be intelligent life out there, but it would be a top-of-the-food-chain predator like man—or like the aliens in H. G. Wells's *The War of the Worlds*. As Poe had prophesied for a more fully evolved mankind, Wells's aliens are "heads—merely heads," and like Wells's proletarian Morlocks and so many other aliens in later SF, they ate people—or, more accurately, "they took the fresh, living blood of other creatures, and *injected* it into their own veins." Wells's narrator comments, "The bare idea of this is no doubt horribly repulsive to us, but at the same time I think that we should remember how repulsive our carnivorous habits would seem to an intelligent rabbit."[3]

The irony and "cultural relativism" of that passage will be a common feature of much SF that deals with aliens, even in the most naive pulp stories. For if we may enter battle with a cry like E. E. Smith's Skylark of Valeron—"Humanity *über alles!*"—why shouldn't aliens adopt the same attitude. Wells puts a spin on this ethical theme by likening his aliens to the technologically superior forces of European imperialists' asserting their dominion over the "savages" of what had yet to be called the Third World. On the Earth to be colonized by Wells's Martians, not everyone will be eaten, at least not all at once. The human race will be domesticated, or so the "artilleryman," Wells's most survivalist-minded character, imagines:

> "Very likely these Martians will make pets of some of them, train them to do tricks—who knows?—get sentimental over the pet boy who grew up and had to be killed. And some, maybe, they will train to hunt us."
>
> "No," I cried, "that's impossible! No human being—"
>
> "What's the good of going on with such lies?" said the artilleryman. "There's men who'd do it cheerful."[4]

I can't resist a flash forward to a recent space opera that may represent the Soldier of Fortune school of SF in its final (so far) decadence.

William Barton's *When Heaven Fell* (1995) takes the artilleryman's hint as its premise and portrays a band of conscript humans fighting dirty wars on behalf of a race of alien lizards. The march tempo is the same as in the mercenary fantasies of Pournelle, Drake et al. Only the flag has changed. A similar premise informs the highly accomplished sequence of novels by Iain M. Banks that begins with *Consider Phlebas* (1987). In both series, the plotline promises to come down eventually on the side of a human imperium, but less in the spirit of E. E. Smith than in that of A. E. Housman, in his poem, "Epitaph on an Army of Mercenaries," which Barton quotes as his epigraph, and which I cannot resist quoting in turn, it so well encapsulates our theme:

These in the day when heaven was falling,
  The hours when earth's foundation fled,
Followed their mercenary calling
  And took their wages and are dead.

Their shoulders held the sky suspended;
  They stood and earth's foundations stay;
When God abandoned, they defended,
  And saved the sum of things for pay.[5]

Housman, of course, did not imagine the mercenaries' paymaster to be aliens, except in the sense that an imperialist force is alien to the territories it subjugates. But the moral remains the same—in Wells, in Banks, in Barton.

The image of the alien, like that of the robot (discussed in the Introduction), has many other valences than that of a conquered race or its conqueror. Besides invading aliens and their occupying forces, there are also, in a friendlier but still problematic way, resident aliens. That lesser breed before the law has been much to the forefront these days in both national politics and SF. Perhaps because they have the same name, *alien*, the extraterrestrials of traditional SF have been easily assimilated into the traditional workforce of those recently crossing the border into the United States, legally and illegally. The Coneheads of Remulak, who made regular appearances on *Saturday Night Live*, have been the most memorable prime-time aliens, lovable but clueless klutzes in the venerable tradition of immigrant-stooge comedy, which may go back to the first Native American to do a stand-up impression of the Pilgrim fathers after Thanksgiving dinner.

The movie and then television series that pursued this metaphor most systematically was *Alien Nation* (film, 1988; series, 1989–1990), in which a quarter-million humanoid aliens have arrived in Los Angeles as shipwrecked refugees. Having been bred and trained on their home planet to be slaves, they have already been assimilated into the workforce as the series begins. Like the aliens on *Star Trek* or *Babylon Five*, the Newcomers (or "slags," as bigots call them) resemble the other white actors on these shows, except for minor cosmetic differences. This may be dictated by budgetary necessity, but it does underline the show's liberal message: that aliens are people very much like us, entitled to have their fair share of the American pie. There are some multicultural differences, of course: the Newcomers have some peculiar food preferences (curdled milk and organ meats), and their taste in home furnishings can seem garish. But they have adapted to their new home to that degree that many have become Roman Catholics. It's not hard to figure out who the Newcomers are on the basis of these and other clues: they are Chicanos.

As such they are the lineal descendants of Swedenborg's aliens, Native-Americans once-removed. This equation of aliens with Third World denizens has become such a commonplace that a recent political cartoon by Toles shows one scientist looking into a microscope and exclaiming to another, "Life on Mars! Do you realize what this could mean?" The answer to that appears in the panel below, where a pair of tiny stalk-eyed Martians are entering a sneaker factory on Mars. One Martian to the other: "So . . . Is 3¢ a year a good wage or isn't it?" Another Martian down in the corner comments, "They're warning that the Venetians will do it for 2."[6]

This is not to say that all extraterrestrials in SF stories are earthly aliens dressed for Halloween. Like the robot, the alien is a disguise that can be adapted to many purposes. Anyone perceived as strangely Other can be presented in alien disguise. In 1958 in that immortal turkey, *Queen of Outer Space*, starring Zsa-Zsa Gabor in her defining role, voyagers to Venus discovered an alien race of beautiful, man-hating women whom the voyagers converted to the ways of Earth; any number of B movies of the '50s and '60s discovered similar aliens in bathing suits and body stockings. Aliens might also be Communists, as in Heinlein's *The Puppet-Masters*. Or they may be the embodiment of Reason and Science, like Dr. Spock or the Luciferian Overlords of Arthur C. Clarke's *Childhood's End*. They may come visiting as gods or angels, like Clarke's transcendental Overmind in that same novel, or like the aliens in Steven Spielberg's smash-hit religious allegories, *Close Encounters of the Third Kind* and *E.T: The Extraterrestrial*. The one thing they cannot easily be is Us (except for purposes of parody: the Coneheads, in their imperfect understanding of consumer culture, are Us and then some).

As in classic psychoanalytic theory, the opposite is usually true. In the distorting mirror of fantasy, Aliens are Us by the same dark logic that equates Dr. Jekyll and Mr. Hyde. In the marketing of pop culture, there is little distinction between monsters of supernatural horror and monsters from outer space: Dracula, King Kong, the Creature from the Black Lagoon, the denizens of *Jurassic Park*, and the Geiger-inspired aliens of *Alien* all derive their scariness not so much from their fangs and feral natures as from their resemblance to that scariest monster of all, the Being in the Black Mirror.

That Being, for all the ways he may be disguised, has a limited emotive range. His two aims in life are rape and murder, or some combination thereof—hence, the icon on so many ancient SF magazines of the virtuous maiden being threatened by the Venusian invader's octopus arms. I speak of that Being as "he" advisedly, since the Being in the Black Mirror has traditionally been a male figure invented by a male writer for an audience of male readers, mostly in their teens. Women who ventured into this realm rarely lingered. The experience was too much like going into the grungiest of topless bars. Yet women who did stay to watch the show, and to reflect on it, and eventually to write their own tales of aliens, were inspired to reengineer the old trope of Alien = Opposite Sex into something quite different.

Ursula Le Guin's first large success, *The Left Hand of Darkness*, depicts androgynous aliens who impart a lesson in equal opportunity to a visiting human male. The difficulty with Le Guin's Gethenians, as with most of her other aliens, is that they seem to spring with such pure didactic intent from the Swedenborg tradition of populating outer space with noble savages of ideal virtue. Indeed, in *Always Coming Home* (1985), she creates a postindustrial California that has reverted to an ideal pre-Columbian condition, which is, of course, matriarchal. She no more endeavors to account for how this came about than Swedenborg thinks to ask how the inhabitants of Jupiter and Saturn came to be Christians. It's enough in both cases that things are as they should be.

A more audacious twining of the two antitheses of alien-human, male/female is to be found in James Tiptree, Jr.'s "The Women Men Don't See" (1973). Tiptree herself embodied the dichotomy she so often wrote about, having published under a male pseudonym for the first decade of her career in SF (1967–1977), and thereby lured a few SF grandees (notably Robert Silverberg) into categorical declarations that "Tiptree" had to be a male writer—despite the evidence of stories like "The Women Men Don't See." In that tale, two middle-class American women, mother and daughter, vacation in the Yucatan with two purposes in mind: for the daughter, to become impregnated by the Mayan-descended Esteban, captain of their charter plane, and then for both women to rendezvous with aliens who can take them away from the human society in which they feel themselves to be an alien species. "What women do," the mother explains,

> "is survive. We live by ones and twos in the chinks of your world-machine."
>
> "Sounds like a guerrilla operation," [the narrator comments].
>
> "Guerrillas have something to hope for." Suddenly she switches on the jolly smile. "Think of us as opossums, Don. Did you know there are opossums living all over? Even in New York City."
>
> ". . . "Men and women aren't different species, Ruth. Women do everything men do."
>
> "Do they?" Our eyes met, but she seems to be seeing ghosts between us in the rain. She mutters something that could be "My Lai" and looks away. "All the endless wars . . ." Her voice is a whisper. "All the huge authoritarian organizations for doing unreal things. Men live to struggle against each other; we're just part of the battlefield. It'll never change unless you change the whole world. I dream sometimes of—of going away—"[7]

In the event Ruth and her daughter are rescued *by* the aliens *from* the Earthmen, and if the aliens are not full-scale phallomorphic bug-eyed monsters, they come close enough: "They are tall and white. . . . The one nearest the bank is stretching out a long white arm towards Ruth. She jerks and scuttles farther away. The arms stretches after her. It stretches and stretches. It stretches two yards and stays hanging in air. Small black things are wiggling from its tip." A fate worse than death? No, for Ruth and her daughter, that's what *men* are. The aliens are a ticket to freedom.

Turnabout is fair play. What's sauce for the goose is sauce for the gander. Do unto other as they did unto you. These are the karmic laws of all pulp fiction. One of my earliest memories of this paradigm with respect to alien invaders was a story in an old E.C. comic book. An avid fisherman sees a candy bar, takes off the wrapper, and finds that the candy bar is bait, and now, as he's reeled into a flying saucer, he's the fish. Tiptree's story is more complicated in its system of reversals, but the final line of her story is not unlike the comic book's moral: "Fish are to men, as men to aliens: they kill us for their sport." Tiptree concludes, "Two of our opossums are missing."

I think it is telling that Tiptree chose the Yucatan as her setting and parallels the mother's tryst with the aliens and the daughter's one-night stand with the Mayan Esteban; that Le Guin's template for utopia should look back to the Indian tribes of North America; even that the Newcomers of *Alien Nation* evidence so many parallels to Mexican immigrants. The reason should be evident: the most well-documented collision between alien cultures took place on this continent as the nations of Europe made "first contact" with the various indigenous cultures, decimated their populations by plague and warfare, and then enslaved and imprisoned the remnant of survivors. Accomplished over the course of centuries during which it became ever more difficult to whitewash the genocidal record with agreeable foundation legends, the tragic fate of North America's indigenes would haunt American writers from Cooper through Faulkner and on. The history of that haunting has been well chronicled by Leslie Fiedler in *Love and Death in the American Novel* (1960), and I won't attempt to paraphrase Fiedler's argument here, except to note that SF has been a significant perpetuator of that quintessential American tradition.

Earlier generations of SF writers may have written on such themes as alien invasions and first contact without reflecting on the historical parallels to their tales, but such innocence is to be found now only in work executed at the moron level of Ed Wood, creator of *Plan Nine from Outer Space*, in which the invading aliens are just plain folks dressed for Halloween. There can still be easy oppositions of Good Guys and Bad Guys, Us and Them, but the codes have been broken. Here, in an instant analysis of American culture provoked by the release of *Independence Day*, the critic James Bowman connects Fiedler's theme with the latest wrinkle in SF:

> The contemporary American imagination . . . has raised its sights above the terrestrial horizon to the shores of space. Thence are expected to come new cannibals or noble savages to mark out a conception of otherness not unlike that of the fifteenth century and what Rawson calls its "division of natives into good and bad savages, whose prototype is Columbus's view of Arawaks and Caribs."
>
> Just now, it is the bad savages that most interest Americans. . . . Now forgotten are such paeans to the good savage as *Close Encounters of the Third Kind* and *E.T.* It is as if *Stagecoach* or *Northwest Passage* had been remade without any acknowledgement that Hollywood has spent most of the intervening decades teaching us to love and pity the Red Indian instead of shooting him.[8]

The question of whether the pre-Columbian cultures of North American were good or bad is not entirely academic, but the moral self-image of America must rest in large part on whether the settlement of the continent by white Europeans was an act of usurpation and genocide. Even as an academic question, the issue has lately been a source of active controversy, as revisionist historians unearth mass graves of data from the libraries and their conservative opponents denounce them as desecrators of the flag. If the revisionists are right, then we are as a nation steeped in a blood guilt that has never been expiated because never confessed, and every acre of land bears its own ancient curse. Of course, over the centuries, history has drenched pretty much the entire globe with crimes of expropriation, so America would not be unique.

But it has always *wanted* to be unique: a new Chosen People with God's own seal of approval. That's why the controversy has become so heated. If the liberals were right, then the Indians still have a moral claim on those lands that they were, by various means, cheated of, a claim that might encompass the whole continent. But if the continent is the white man's by virtue of his Manifest Destiny, then all of outer space belongs to him as well. The two claims—retroactively on the continent; prospectively on the galaxy—issue from a single sense of moral entitlement.

No other writer in the SF field has been drawn to this theme so frequently or to such striking effect as Orson Scott Card. Card is a relatively late arrival on the SF scene, but since his first novel, *Ender's Game* (1985), Card has won numerous awards, produced a body of work as remarkable for its quantity as its popularity, and may well be considered Heinlein's true heir—not because his fictions recirculate the Master's mannerisms and ideology (as does the work of Pournelle & Co.) but because of his knack for attack-dog narrative, his fertile invention, and the moral conviction he brings to his work. What he also brings to his work is Mormonism, the faith he was reared in and whose tenets and history have provided the allegorical subtext to so much of his work.

Mormons have a unique perspective on American history. In their enforced migration westward, they are a communal embodiment of Manifest Destiny. Yet their attitude toward the Indian tribes whose orbits intersected theirs was not antagonistic but cautiously respectful, for according to *The Book of Mormon* the indigenes of North America were descended from a lost tribe of Israel, the Nephites, who voyaged to North America with miraculous aid, circa 200 B.C., and were later visited by the resurrected Christ. The early Mormons actively sought to convert Indians/Nephites to Mormonism (while refusing until very recently to welcome blacks into their fold), and their missionaries are still most active in Central and South America. Card spent his required two years of missionary labor in Brazil during the time of the Vietnam War. It is hard to think of a more ideal apprenticeship into Otherness for someone aspiring to a career in SF.

Ender's Game, based on Card's first published SF story of the same title (1977), conflates a number of tried-and-true SF elements into a plot that, like much of Heinlein, would seem steeped in irony if it were not presented in a tone so dead earnest. The protagonist, Ender Wiggen, is that most beloved figure of the genre, a child genius of modest circumstances destined to be the savior of his race while still an adolescent. Not only that, but his two siblings are almost as remarkable: they become the intellectual leaders of their time while they are still children. This one-ups Theodore Sturgeon's *More Than Human*, but Card further flatters his aimed-for constituency of bright juveniles by having Ender's special talent be a genius for arcade games. The military, discovering Ender's pinball wizardry, recruits him into a special corps of youthful arcade-gaming geniuses, and Ender, thinking he is zapping only computer-simulated enemies, saves the human race from destruction by destroying a galactic civilization of Heinlein-like insect enemies. Heinlein called them Bugs; Card, archly, calls them Buggers, to make them doubly objectionable.

It is, then, quite at the end of his novel that Card pulls off a memorable *coup de théâtre:* the Buggers aren't really bad. They just didn't realize that the humans they'd attacked were a form of intelligent life like themselves. The child Ender has been the unwitting instrument of an unmerited genocide, and so, in the sequel, *Speaker for the Dead* (1986), the mature Ender tours the galaxy confessing his sin and trying to find a suitable home for the one surviving Bugger queen whom he has rescued from his one-man holocaust (not the only such event in Card's oeuvre). Once more, he discovers an alien race, the Pequininoes of the planet Lusitania, who seems to be bad (they crucify a missionary) until they're shown, like the Buggers, to have been misunderstood. Another sequel, *Xenocide* (1991), does not fully resolve all the complications, but its title does signal Card's moral and allegorical concern: how can one hope to live a moral life as a citizen of a nation drenched in the guilt of unjust and genocidal wars?

In a later and still ongoing series of novels, *The Tales of Alvin Maker,* Card poses the same question, more artfully, in a saga that transmutes American—and Mormon—history into the stuff of legend.[9] The series posits an alternative Federal Age America in which the Revolution fizzled and there is a Macedonian salad of smaller states and alliances; it also employs that staple trope of lazy fantasists: a world in which *magic really works.* The hero, once again, is a preternaturally gifted boy, Alvin Maker (aka Alvin Smith), destined to become a world savior. Look at the pattern of the plot in one light, and Alvin is a stand-in for Joseph Smith, the founder of Mormonism, but look at it in another way, and he is Wagner's Siegfried, an apprentice blacksmith like Alvin.

At this point in the story, with four volumes in print, the parallels of Alvin's and Joseph Smith's careers have only begun to be developed, though Alvin's world, with its supernatural embroideries of folk magic, would seem to be one that Joseph would have no trouble believing in. The Wagnerian parallels are more telling. Neither Card nor any of his critics whom I've read have noted how his plot, especially in the latter two books, derives from Wagner's *Ring* cycle, sometimes on a scene-for-scene basis. The truth-telling woodland bird makes an appearance, as do the fate-weaving Norns (who deliver great chunks of exposition, as they did in *Gotterdammerung*), and the heroine is a dead ringer for Brunhilde, a woman who superintends the hero's childhood from afar and is destined to be his consort.

It is artfully done, with interesting reversals of Wagner's moral equations. Thus, Alvin is brought up, like Siegfried, by a foster father of an alien race—but not a scheming Nibelung like Mime; rather, a noble Red Indian, Tenskwa-Tawa, the "Red Prophet" of the second book in the series. By this association, Alvin is empowered, like Cooper's Deerslayer, with the secret wisdom of the Red Indians and their supernal strengths. "He could run like a Red man," Card writes:

> Just like a Red man, that was how he moved. And pretty soon his White man's clothing chafed on him, and he stopped and took it off, stuffed it into the pack on his back, and then ran naked as a jaybird, feeling the leaves of the bushes against his body. Soon he was caught in the rhythm of his own running, forgetting anything about his own body, just part of the living forest, moving onward, faster and stronger, not eating, not drinking. Like a Red man, who could run forever through the deep forest, never needing rest, covering hundreds of miles in a single day.[10]

The Alvin Maker series plays games with American history that sometimes improve it in wish-fulfilling ways. His red men have held their own against white men in a way that bodes well for their own (alternative) future, and through Alvin's intercession the curse that had been placed on the people of Vigor Church, Alvin's hometown, is lifted. The men of Vigor Church had been responsible for a massacre of Indians at Tippy-Canoe, since when they were under a compulsion to confess their crime to any stranger they encountered; nothing less than full confession would staunch the bleeding of their guilty hands. The shaman Tenskwa-Tawa alone can remove the curse. Will he do so, so that the latter-day saints of Vigor Church can inhabit the Far West? Alwin implores:

> "I didn't see my answer," said Tenskwa-Tawa. "So I have had all these months to think of what it was. In all these months, my people who died on the grassy slope have walked before my eyes in my sleep. I have seen their blood again and again flow down the grass and turn the Tippy-Canoe Creek red. I have seen the faces of the children and babies. . . . Each one I see in the dream, I ask them, Do you forgive these White murderers? Do you understand their rage, and will you let me take the blood from their hands? . . .
>
> "If even one of them had said, I do not forgive them, do not lift the curse, then I would not lift the curse. . . .
>
> "I also asked the living. . . . Those who lost father and mother, brother and sister, uncle and aunt, child and friend. . . . If even one of these living ones had said, I cannot forgive them yet, Tenskwa-Tawa, I would not lift the curse."
>
> Then he fell silent one last time. This time the silence lasted and lasted. . . . Like Alvin, he, too, had wept, and then had waited long enough for the tears to dry, and then had wept again. . . . Now he opened his mouth and spoke again."I have lifted the curse," he said.[11]

Abolition remains an ungoing battle as the fourth volume comes to an end, but Card probably intends a redemptive ending for the country at large, as well as for Alvin's band of latter-day psychics. Alvin's own fate will be a test of the author's willingness to defy the genre's resistance to tragic endings, since by both templates he has adopted (Joseph Smith and Siegfried), Alvin would seem destined for tragedy. However, he has magical powers that Joseph Smith and even Siegfried lacked, so perhaps he will solve all the basic problems of American history and still get the girl, the gold watch, and everything. The remarkable thing is that Card's flair for storytelling is such that even sophisticated readers can be engaged by his inventions. He is the Edgar Rice Burroughs of Generation X.[12]

Science fiction has a rather narrow shelf of books that deal, directly or allegorically, with the other "alien" presence that has informed American history. A tribe of nightmarish blacks appear toward the end of Poe's *Narrative of Arthur Gordon Pym*, and there are other dismally caricatured black characters in his shorter tales. Pulp fiction of the nineteenth century regularly casts blacks as subsidiary villains. A sample of such writing at its most disgraceful can be found in George Lippard's *The Quaker City* of 1844, the best-selling American novel before *Uncle Tom's Cabin*. In the futures depicted in the earliest SF of the pulp magazines, blacks are as scarce as they would have been on Baffin Island, except insofar as SF's penchant for lustful aliens may derive from similar fantasies of interracial rape (e.g., Griffith's *Birth of a Nation*). This situation persisted well into the '50s and '60s. *The Encyclopedia of Science Fiction*, which has articles on SF's treatment of just about everything else under the sun, has no single entry on race, an absence that is partly explained by this remark in its entry on "Politics": "Direct treatment [of racial themes] seemed too sensitive to most genre-magazine editors, who preferred their writers to use aliens in parables whose arguments were conducted at a more abstract level. . . . In general . . . as the real-world problems become ever more urgent, the tendency of genre SF has been to ignore the issue or sanctimoniously to take for granted its eventual disappearance."[13]

This is not so. On the one hand, one award-winning black SF writer, Octavia Butler, has made racial (and sexual) confrontations, in the future and in alternative pasts, almost her exclusive theme. Like Card, she works on large canvases. A first series of *Patternist* novels ran to five volumes, and that was followed by the *Xenogenesis* trilogy. Both sequences concern the interracial, or interspecies, breeding of humans to improve the species—in the first case, to create mutants with psychic powers; in the second, to defuse human aggressiveness. Butler's attitude to these undertakings is interestingly ambivalent. She has a New Age enthusiasm for telepathy, out-of-body experience, and kindred knacks, but her eugenics programs are run by malign supernatural beings or by Strieber-style alien experimenters. Her black heroines must endure rape, incest, and slavery in the Old South, all for the sake of improving the race. As with Card, or the gothic fantasist Anne Rice, an intense conviction coupled with a total lack of humor allows Butler to invent compelling, if implausible, plots. The moral that usually seems to emerge—that miscegenation is a good thing, albeit very unpleasant—is one more likely to appeal to readers of John Norman than those of Ursula Le Guin, but that is any author's prerogative. As SF's only prominent black writer who has chosen to focus on racial concerns, Butler is not about to be challenged for being politically incorrect.

The same cannot be said for Robert Heinlein, whose 1964 *Farnham's Freehold* is the most reprehended work of a writer who has been much reprehended. The Baen reprint of 1994 even brags about it on the front cover: "Science Fiction's Most Controversial Novel." In Chapter 8, I quoted Hugh Farnham's enthusiasm for the evolutionary potential of nuclear wars, but that strand of the book's plot was not the source of its scandal—which is its vivid representation of the trans-African nation that has developed two thousand years after the nuclear destruction of all civilizations in the Northern Hemisphere. Initially, it might seem rather utopian, at least from a black perspective, for the world has been turned upside down, and blacks are in charge, while the remaining whites are their slaves. Bred to be short and trained to be servile (males are castrated after a short season as studs, and their thumbs are amputated as a further mark of servility), this is the ultimate white dystopia. White women, as per John Norman's Gor novels, are all "sluts," the sexual slaves of their black masters. The chief of these is Ponce, the affable, well-educated Lord Protector of the Noonday Region and the owner of the Farnham family and their friends, who have been bounced into this white supremacist nightmare by the nuclear explosion that kicks the novel off.

Hugh Farnham, Heinlein's hero and ideological spokesman, is a Figaro among white slaves, averting castration and thumblessness and, by initiating Ponce into the pleasures of contract bridge, becoming the Lord Protector's most valued possession. For many chapters, Heinlein seems to give the black devils their due. The tit-for-tat correspondence of white slavery to that endured by blacks in America is worked out with mathematical rigor. Hugh, like Ponce, takes a pragmatic view of history as a narrative written by the winners. He reflects:

> This matter of racial difference—or the nonsense notion of "racial equality"—had never been examined scientifically; there was too much emotion on both sides. Nobody *wanted* honest data.
>
> Hugh recalled an area of Pernambuco he had seen while in the Navy, a place where rich plantation owners, dignified, polished, educated in France, were black, while their servants and field hands—giggling, shuffling, shiftless knuckleheads "obviously" incapable of better things—were mostly white men. He had stopped telling this anecdote in the States; it was never really believed and it was almost always resented—even by whites who made a big thing of how anxious they were to "help the American Negro improve himself." Hugh had formed the opinion that almost all of those bleeding hearts wanted the Negro's lot improved until he was *almost* as high as their own—and no longer on their consciences—but the idea that the tables could ever be turned was one they rejected emotionally.[14]

The shocking truth, gradually revealed, is that Ponce and all his minions are cannibals, who regularly dine on the well-prepared flesh of their slaves. Ponce prefers young white girls, but never those of his own breeding. Heinlein affects to excuse this on terms of cultural relativism. "Sweetheart," Hugh explains to his mistress, "don't hold what he ate too much against Ponce. He honestly did not know it was wrong—and no doubt cows would feel the same way about us, if they knew." Worse than the unknowing Ponce is Farnham's own servant, Joe, who had been allowed admittance to the fallout shelter and thus been transported with the whole family into this cannibal future. Joe not only dines on white flesh, but he intends to satisfy his lusts on Farnham's wife and his castrated son. When Farnham takes exception to Joe's intentions, Joe replies, "The shoe is on the other foot, that's all—and high time. I used to be a servant, now I'm a respected businessman." So much for liberalism. Once the shoe is on the other foot, blacks become cannibals and reenact the rituals of rape and murder from *Birth of a Nation*.

Farnham's Freehold appeared in the mid-'60s at the height of the civil rights movement and must be considered a response to those events. Heinlein in earlier books had scrupled to appear tolerant in matters of race. The astronaut hero of one of his young adult SF novels turned out to be, quite *en passant*, a black. Even in *Farnham's Freehold*, the hero upbraids his son for derogatory remarks about "niggers." For all that, Heinlein is clearly choosing sides in this novel. He is endorsing D. W. Griffith and spurning Martin Luther King, and such is the energy released by this surrender to Id forces and maximal political incorrectness that the book is one of Heinlein's zestiest. He enjoys being scandalously Bad, like William Burroughs or Janis Joplin, and even if you don't approve, the enjoyment of witnessing a taboo artfully broken is contagious.

Just as Heinlein had been among the first of the old-guard SF writers to take advantage of the new erotic candor, so was he a pioneer in opening the Pandora's box of overt racism. Without the example of *Farnham's Freehold*, publishers might have been less willing to the push the envelope of the genre by publishing the various survivalist series cited in the previous chapter. Pournelle and Niven might have had second thoughts about the passages in their 1977 catastrophe novel, *Lucifer's Hammer*, in which blacks in the United States revert to cannibalism not in the two thousand years Heinlein allows but in something closer to two months following Earth's collision with a comet.

Realistically, neither collisions with comets nor black cannibals are anxieties to be entertained with any urgency. Rather, they are like the millenarian hopes of those believers who like to imagine their enemies confounded by God's Judgment. From their celestial grandstand where the raptured hang suspended, they may watch with delight as the unrighteous perish, and if anyone takes exception to such an entertainment as salacious or pornographic, the authors will insist it's all just escapist sci-fi fun.

Is it? Here is one of the extremer examples of such sci-fi fun from a novel that appeared in 1978, a year after *Lucifer's Hammer*, and one that also features the sudden reversion of L.A.-area blacks to cannibalism:

> The night is filled with silent horrors; from tens of thousands of lampposts, power poles and trees throughout this vast metropolitan area [Los Angeles] the grisly forms hang.

> In the lighted areas one sees them everywhere. Even the street signs at intersections have been pressed into service, and at practically every street corner I passed this evening on my way to HQ there was a dangling corpse, four at every intersection. Hanging from a single overpass only about a mile from here is a group of about thirty, each with an identical placard around its neck bearing the printed legend, "I betrayed my race." Two or three of the group had been decked out in academic robes before they were strung up, and the whole batch are apparently faculty members from the nearby UCLA campus.

Can you guess the author? It is not Jerry Pournelle, though from the L.A. locale and the author's animus against academia, that would be a reasonable guess. No, it's from the *Turner Diaries* by William Pierce, a novel of only modest infamy until it was reported to have been held in reverence by Timothy McVeigh, who was convicted of blowing up the Oklahoma City Federal Building in 1995. Another example of life imitating science fiction—for such bombings are commended and diagrammed in *The Turner Diaries*. For some years McVeigh acted as the book's evangelist, traveling to gun shows and selling copies (not then available at bookstores) at five bucks a shot.

Pierce's plot traces the arc of a standard action-adventure novel, with a gradually escalating series of acts of terrorist violence. The difference is that Pierce's hero, Earl Turner, is the perpetrator, an arch-terrorist who chronicles his criminal career in the matter-of-fact tones of a high-tech Robinson Crusoe. One early vignette (page 10) depicts the murders of a black liquor store clerk and two Jewish deli owners. Seven pages later, a neo-Nazi's death is avenged when the Cook County sheriff has his head blown off by a shotgun. When a "responsible conservative" condemns such behavior on TV, his car is bombed. The ante is upped steadily. Bombs level buildings; the U.S. Capitol is attacked by mortars. When the Organization, as Turner's group calls itself, gets hold of nuclear warheads, whole cities are wasted—even in the Epilog, the entire continent of Asia. It is all in the interest of achieving racial purity across the face of the planet.

Pierce's object as a novelist is to create a plausible scenario emphasizing weaponry and the logistics of its deployment, a kind of how-to book for mad bombers. People, in the usual novelistic sense, scarcely enter into it. Earl has a girlfriend, Katherine, by way of establishing his bonafides as an action hero, but Pierce doesn't waste his time on character, dialogue, or mere interpersonal conflict. He is just as indifferent to creating realistic portraits of enemies, even of the two-dimensional variety one may find in Pournelle's work. His bad guys rarely survive for more than a single page.

Bad guys are legion. Basically everyone who is not an official member of the Organization is an enemy of the white race and so deserving of death. Among those strung up on the Day of the Rope (described in the quote above) are "the politicians, the TV newscasters, the newspaper reporters and editors, the judges, the teachers, the school officials, the 'civic leaders,' the bureaucrats, and all the others [who implemented] the System's racial program"—in short, the entire upper-middle class, all professions, and government employees at all levels. Pierce devotes a few passages of stern moral instruction to explaining why thousands and millions deserve to die, and the reasons he gives will be familiar to all true believers: the future requires it. The writers of the School of Heinlein would agree in principle; they differ only in whom they have slated for destruction.

When I first wrote about *The Turner Diaries* in an article commissioned and then scrapped by the *Nation*, the book had been much deplored and its purpler passages quoted in news stories, but it could not be obtained at ordinary bookstores. Because of the Oklahoma City bombing and the attention the book had received (a front-page story, with many quotations, in the *New York Times* of July 5, 1995), I predicted that someone in the publishing industry would find such free publicity irresistible and bring out the once self-published book in a mass-market edition. My prediction proved true.

For all that, I remain a First Amendment absolutist, but on aesthetic more than political grounds. The fact is that Pierce is such an incapable artist that his book can preach only to the converted. Timothy McVeigh probably did not need the incitement of Pierce's novel to commit his crime; he was heading there already. Other as-yet-uncoverted readers would only be likely to see in Pierce's pages the impress of an irrational hatred, one that simply cannot *imagine*, much less portray, the Other. By contrast, Heinlein, Pournelle, and even Newt Gingrich know they should make an effort to enter into the consciousness of even those characters they most deplore. Heinlein, as the better artist, is more successful, and his Ponce almost overcomes his handicap as a designated villain. Capable storytellers tend to have memorable villains; even when their deeds are vile, their personae are vivid and true.

I do not mean to suggest that all aliens in SF are to be deciphered as bad guys of other ethnic backgrounds or the opposite sex in disguise. The aliens in movies have tended to come in one or the other of those flavors, usually for demographic and budgetary reasons. Demographically, simple tales of Us and Them appeal to the lowest common denominator. Budgetarily, people in costumes are a whole lot cheaper than any of the available alternatives. And people in costumes (whether with green skin, warts, funny ears, extra eyes, fur, or talons) register as at least multicultural and usually sinister.

The aliens in books are another matter, often. So much of the best SF about aliens has never made it to the big screen because only their prose skills limit the special effects that writers can lavish on their aliens. Any number of writers and directors were stymied by the problem of translating Frank Herbert's *Dune* to the screen, and when David Lynch's lame attempt appeared in 1984, one had to admire the wisdom of those who admitted defeat before spending millions.

The problem was not just the hokeyness of the gigantic "sandworms" and the unpersuasive extraterrestrial settings. The story was incoherent. Screenwriters do not have the luxury that novelists enjoy of taking the time to explain things, to pose riddles and then work them out, to *think*. Such bemusements can be the glory of SF (as of the deductive mystery, another genre poorly served by film), and it is a glory as accessible to younger readers as to adults, provided they have some background in basic science (a more likely proviso for bright twelve-year-olds than for most adults). When I first read Hal Clement's *Mission of Gravity,* which ran as an *Astounding* serial in 1953, I thought it the best account of alien life on another planet that I'd ever read. Forty-three years later, my opinion has not changed. It may be that my response is conditioned by that earlier first contact; I would not be the first SF critic to mistake the redolence of childhood memories for the music of the spheres. But even if there are now rival claimants to "best," it surely set the standard for how the task is to be judged.

What distinguishes *Mission of Gravity* is the author's zeal for the scientific process. Hal Clement (whose real name is Harry Stubbs) was a high school science teacher with degrees in astronomy, chemistry, and education. On the evidence of his SF, he must have been a Merlin among science teachers, for he is able to turn Newton's laws of motion and the periodic tables into matter fit for a saga. Space operas like *Star Trek* make obeisance to science, but the words of the script rarely signify. They only need to *sound* like a science classroom, as when voyagers to Venus in *Queen of Outer Space* explain why they can leave their rocket ship unhelmeted: "The gravity's so close to Earth's that the atmosphere must be breathable." SF in written form rarely gets as dumb as that, and it regularly is dense enough as to defy the easy comprehension of those on the other side of the two-culture gap. *Dense*, in such cases, is not necessarily a pejorative. *Dense* can also describe pecan pie.

In an article, "Whirligig World," that appeared in *Astounding* after *Mission of Gravity* had run as a serial, Clement explained how he developed the specifications for his aliens from the peculiar characteristics of their planet Mesklin, which orbits about a known star in the 61 Cygni system. Sixteen times as massive as Jupiter, with a diameter equal to Neptune's, Mesklin's surface gravity is three hundred times what we're used to on Earth. However, because of its rapid rotation (the Mesklin "day" is 17 3/4 minutes long) and the consequent centrifugal force generated, Mesklin has come to have the shape of a fried egg rather than a sphere, and so gravitational forces can vary from 3 times that of earth at its equators to 665 times at its poles.

Clement goes on to explain how these characteristics would be likely to determine the planet's weather systems and then to speculate as to what the chemical basis for life would be on a planet where ammonia would melt on the very hottest days. He opts for methane as the best substitute for water, but even methane has its problems, for at Mesklin's probable atmospheric pressure methane boils, so that

> one sixth of the year, the planet will be in a position where its sun could reasonably be expected to boil its oceans. What to do?

> Well, Earth's mean temperature is above the melting point of water, but considerable areas of our planet are permanently frozen. There is no reason why I can't use the same effects for 61C; it is an observed fact that the axis of rotation of a planet can be oriented so that the equatorial and orbital planes do not coincide. I chose for story purposes to incline them at an angle of twenty-eight degrees, in such a direction that the northern hemisphere's midsummer occurs when the world is closest to its sun. This means that a large part of the northern hemisphere will receive no sunlight for fully three quarters of the year, and should in consequence develop a very respectable cap of frozen methane at the expense of the oceans in the other hemisphere. As the world approaches its sun the livable southern hemisphere is protected by the bulk of the planet from its deadly heat output; the star's energy is expended in boiling off the north polar "ice" cap. Tremendous storms rage across the equator carrying air and methane vapor at a temperature little if any above the boiling point of the latter; and while the southern regions will warm up during their winter, they should not become unendurable for creatures with liquid methane in their tissues.[15]

If your response to such a passage is for your eyes to glaze and your attention to wander, you had best remain aboard the Starship *Enterprise*, where the "science" component of the SF is limited to the iconography. If, however, you hunkered down to make sense of it, then you might well enjoy the whole novel, whose hero Barlennan is a fifteen-inch, armor-plated, centipede-like alien adapted to cope with the conditions of Mesklin. A sea captain on its methane seas, Barlennan is despatched to retrieve the records of a Terran experimental rocket that has crashed to the surface of Mesklin, whose tremendous gravity prevents its human programmers from doing that job themselves.

The resulting quest is a cross between *The Wizard of Oz* and a class in higher algebra; it is fun, instructive, and *innocent*. No doubt a dedicated Foucauldian might ferret out some objectionable political similitude. The Mesklinites might be seen as alienated, coolie laborers succumbing to the lure of imperialism. Or the absence of women throughout the story might aggrieve feminists. But such exception taking would require a deeply set ideological determination to be blind to what is so clearly the moral of the story: that cleverness, curiosity, fortitude, and cooperation will help anyone, human or alien, to achieve his or its goals.

Sometimes a cigar is just a cigar. Sometimes an alien is just an alien. And sometimes the whole world is mud luscious and puddle wonderful.

Chapter 10

THE FUTURE OF AN ILLUSION— SF BEYOND THE YEAR 2000

Science fiction is an industry, or rather a major component of two large industries, the movies and publishing. The slice of the pie that SF represents in each case is immensely greater than it was only twenty-five years ago, especially for film. In 1971 the overlapping categories of SF, horror, and fantasy accounted for only 5 percent of U.S. box office receipts; in 1982 that figure was approaching 50 percent; in 1990, it was down to 30 percent. More than half of the ten top-grossing films of all time have been SF, including *E.T.* (with $400 million domestically), *Jurassic Park* (over $350 million), and the *Star Wars* and *Indiana Jones* movies. Hollywood's largest budgets, approaching or exceeding $100 million, commonly are lavished on SF movies like *Terminator* 2 and *Waterworld* (each in its day the most expensive movie ever made). Returns can be proportional. *Independence Day,* with a $70 million budget, is edging toward $300 million in domestic receipts.

When the global profit potential of Hollywood films is factored in, the figures become even more impressive. Movies that flop at home can turn big profits abroad. To do so, however, they must appeal to audiences who probably don't speak English and who have widely varying cultural values and intellectual assumptions. This means that dialogue must be kept to a minimum, and action (car chases, fights, explosions) at a max. The optimal ratio is ten minutes of action to two minutes of dialogue, the same rhythm, one commentator has noted, as that of a TV program that must stop and start for ads. This translates into a simple rule of thumb: keep it dumb.

This is actually an easier rule to observe in an SF or horror movie than in movies set in the real world. The SF movies have a long tradition of being dumb, going back to *Godzilla* and *King Kong*. If Hollywood can fill the screen with an authentic-looking dinosaur, who needs any more dialogue than, "Wow, look at that dinosaur!" Recent advances in computer animation and other special effects have been so rapid and eye-boggling that sophisticated audiences can revel along with Pakistani peasants, their disparate disbeliefs equally suspended by the high resolution of the illusion, the speed of the roller coaster.

Here is the take of the commentator referred to above, as he tries to explain the success of *Independence Day* to a highbrow audience:

> Why were these grown-ups [Bob Dole and William Bennett, who both loved the movie] making fools of themselves over a cartoon? It seems to me that they had been caught in a Hollywood trap. For years, American movies presented the spectacle of pulp aspiring to become sophisticated entertainment—the spectacle of what were essentially stock pop formulae being worked up in order to "say" something. Now the trend has completely reversed. Most American movies are sophisticated entertainment striving frantically to become pulp. . . . *Independence Day* . . . and all the other big summer hits . . . want in the worst way to say nothing. For all their cinematographic wizardry, these movies are at the bottom of the generic barrel. And that is where they want to be, flying in underneath the moral radar.[1]

While this strikes me as an often accurate description of the difference between the movies of back when and right now, I don't think it fairly describes the relation of the filmmaker (in either era) to the hipper part of his audience. For instance, Nick Lowe, the regular movie critic for the British SF magazine *Interzone*, wrote a rave for *Independence Day* that opens with this note of high praise: "Perhaps the boldest of many daringly old-fashioned things about *Independence Day* is its refusal to offer any real platform for merchandising. You might shift a few model kits and console games, but there's nothing you could really stick on a lunchbox, let alone franchise as a range of ubiquitous character dolls (and then repromote on the back of the video release; see the *Toy Story* file for a masterly demonstration)."[2] The rest of Lowe's review is dense with insider lore and wisdom. The reviewing journal *Entertainment Weekly* assumes that its audience has a similar curiosity about the *engines* of the industry and the engineers—the producers, agents, and deal-makers.

No doubt there are still American moviegoers so little used to exercising their imaginations that they can react as Bob Dole did to *Independence Day*—or rather as Louis Menand would have us think he reacted. For Dole could scarcely have given the real reasons he might have enjoyed the movie, beginning with a death scene for Hillary Clinton's doppelganger. Politicians don't have the liberty of saying what they think. Indeed, most people don't. That's why there are movies and novels.

Which brings me to the other aspect of SF as an industry, the book business. Until recently, the sense of the printed word as "product" was confined largely to those in the executive suites of the publishing industry: the publishers themselves, accountants, distributors. Writers tended to think of the machineries of publishing as an unavoidable and cumbersome link between themselves and their public, and their agents and editors encouraged them in this harmless delusion. SF writers, with their active connection to fandom, were especially prone to regard publishing houses as the medium by which their message was delivered to the public rather than as employers exploiting the writers' labor for their own profit.

SF has expanded as a publishing phenomenon over the past five decades but in a much different way from the movies. The movies got bigger and costlier, drawing in adult audiences while keeping a solid grip on the youth market. In published SF, the product has diversified to fill various new marketing niches as they became evident to marketers. The first significant marketing subgenre was that of sword-and-sorcery, which budded off the main body of SF in the mid-'60s, spurred by the success of Ace Books' unauthorized edition of Tolkien's *Lord of the Rings* trilogy. Tolkien imitations were easier to mass-produce and to market than SF, since what readers of sword-and-sorcery wanted was another ride on the same merry-go-round rather than novelty. Sameness is what marketers want us to want.

By the end of the '80s this process of diversification had resulted in at least a dozen "biblio-niches" of SF with relatively little overlap in their readership, each niche offering a sameness all its own. There were, at entry level, niches for comic book readers and readers of *Star Trek* novelizations. There were authors like Piers Anthony and Anne McCaffrey who wrote open-ended series for subteen audiences that were the SF equivalent to the Oz books and to girl-and-horse romances. There were "hard-science" adventures by such heirs of Hal Clement as Greg Benford, David Brin, and Greg Bear; and, distinct from these (but swearing allegiance to the same "hard-SF" banner), the militaristic space operas of Pournelle, Drake et al., and these had their own devolved sub-subgenre of survivalist gore for trailer trash readers. Then there was cyberpunk, which came in at least two forms: the upscale version that placed a premium on conceptual density and the "splatterpunk" variety that offered maximal grossing-out. The writers of these books appeared on the same SF bookshelves in alphabetical order with Ursula Le Guin, Kim Stanley Robinson, Gene Wolfe, and other SF writers of an overtly literary tendency. Indeed, among this lot of writers, as among writers of the mainstream, the process of diversification begins to resemble anarchy, each writer constituting a species to himself or herself.

Writers tend to consider distinction and originality as virtues, but they are anathema to publishers, who value those writers most who can be depended on to turn out product at regular intervals, product that will move through the channels of circulations at a dependable, steady rate. As with fast-food restaurants, a dependable flow generates high profits. Books that are media driven are the likeliest to sell in mass quantities, and so a tie-in to a successful TV series is the best of all possible products. The *Star Trek* franchise has been a gold mine for almost everyone concerned—even, to a degree, for those who have written product for them. Admittedly, their slice of the pie (if any) is smaller than it would be were they to write their own books. At best, they may get a 2 percent royalty, as against standard hardcover royalties of 10 percent and paperback royalties of 6 or 8 percent. But 2 percent of a bundle is better than 10 percent of a smidgeon; a seasoned pro with a good agent can command an advance of $10,000 to $20,000 for a franchised novel, which can be turned out in four or five weeks (or faster). A couple such stints of labor each year will pay the bills and still leave a lot of time for product that isn't "sharecropped" in this way.

What this translates to in terms of the present and likely future of the genre is more of the same and more of the sameness. This is not my opinion only but the received wisdom of those working in the field: editors, agents, and writers. Since it often sounds like grumbling or sour grapes, it is rarely expressed even by those secure in their reputations and incomes, but I recently received a letter from Al Sarrantonio that struck me as having the ring of collective truth. Sarrantonio had been a student of mine in 1974 at the Clarion Science Fiction Workshop at Michigan State University, a summer program that has accrued a remarkable track record over the years for producing SF professionals. Sarrantonio went from Clarion to become an SF editor at Doubleday, which published the largest SF list at that time. He left Doubleday when his own writing career got into high gear; since then he has published over twenty-five books. It's safe to say that he knows whereof he speaks. And here's what he says:

> By 1982 when I got out of the editing business, things had begun to change utterly. Over the previous seven years I had watched my particular company begin to metamorphose from a paternalistic (in the *good* sense) octopus, so big and wonderfully unwieldy that among its eight hundred or so published titles a year there was bound to be something of merit (we tried) toward something gnarled, ingrown, and, by its very nature, anti-literary. The end result of this process was the eventual death of the mid-list. It lives on, barely—or does it? There is evidence that anything not a big seller now is reduced to something less than what the mid-list used to be, which was a training ground. There are more stories: books with three-week shelf-lives, books *out of print*, *pulped* after a matter of months, series books orphaned in mid-stride.
>
> A byproduct of this philosophy (publish *and* perish) has become a "death of history" of sorts, the near-abandonment of the backlist, with the truly frightening result that some of the finest writing in all of SF is now unavailable. What does this bode for the current and next gen-

> eration of SF readers? To be harsh, this is book burning without a match. Who's to blame? Everybody, of course. But it's harder, much harder, to be an editor now than it was in 1982.[3]

Authors who do not write books that conform to the laws of the marketplace don't get published unless they have established a special relationship with a well-placed editor (and there are ever fewer of those, as editors, too, are de-accessioned, to be replaced by younger editors certain to toe the line). And that means that only established authors can write what they would like to. Even that cannot be counted on. Recently, after the death of John Brunner, an English writer noted for his precocity and a prolific output (with an immense backlog of titles to his credit), his estate was valued at significantly less than a thousand pounds. When I knew Brunner, in the '60s and '70s, he had a posh semi-detached house in Hampstead and later an idyllic cottage in far exurbia, where he lived a life of workaholic prosperity. But he did not produce trilogies or even sequels, and could not find takers for the non-SF he wanted to write. At the same time, he grew older, and this has always been a disadvantage in a field whose audience does not, as a rule, age. Authors with a permanent mind-set that is "forever young" have an edge in this regard: Ray Bradbury, Piers Anthony, Michael Moorcock, Roger Zelazny. The list of such perpetually adolescent survivors is extensive.

The list of those who don't survive is more extensive, though the recognition factor is necessarily lower: Brunner, Avram Davidson, Theodore Sturgeon, Alfred Bester, R. A. Lafferty, A. J. Budrys, Robert Sheckley. Like François Villon's roll call of the great dead poets of his day, it is a melancholy assembly. Admittedly, the marketplace cannot be blamed for all human failure. Writers burn out, like Sturgeon, or fizzle out, like Lafferty, or drink themselves to death, like Bester. Even those, like Clarke, Asimov, and Herbert, who managed to extend their commercial success into their emeritus years, have done so by virtue of sheer momentum. Their cachet was great enough that they could recycle their earlier successes into viable product for the industry. Such semiposthumous successes, however, though they may generate sales, rarely register as significant among those readers who look to SF for a handle on the Zeitgeist.

For such readers, there has been only one significant evolutionary event in the field in the past fifteen years: the advent of cyberpunk. The first element of that portmanteau word represents an area in which SF had significantly misfired, by failing to foresee the actual impact that cybernetics would have in daily life. SF had been obsessed with the image of the robot. Ever since Capek's *R.U.R.*, the robot had been a dramatically effective emblem of the possibility that a machine could think, thereby usurping what was supposed to be a human prerogative. The emphasis in SF stories about robots was always on the degree to which they were like humans. Could they be trusted to act as mere appliances, or might they rebel, like HAL in Clarke's *2001?* At what point did their similarity to humankind become identity and entitle them to an autonomous moral existence? Such questions yielded good drama for writers as various as Isaac Asimov, Phil Dick, and John Sladek, but they deflected the genre's writers from noticing the ways in which computer technology was actually reshaping the world.

In his 1982 Pulitzer Prize–winning book, *The Soul of a New Machine,* Tracy Kidder described the development from concept to mass production of a new computer, the Eagle. Along the way he speculates about how the new technology has already reshaped our lives and social landscape, making possible such high-tech wonders as spaceships and junk mail, CAT scans, remote-control weaponry, and advances in meteorology, plasma physics, and mathematics. As to the realm of finance and industry, he wrote:

> Computers probably did not create the growth of conglomerates and multinational corporations, but they certainly have abetted. They make fine tools for the centralization of power. . . . They are handy greed-extenders. Computers performing tasks as prosaic as the calculating of payrolls greatly extend the reach of managers in high positions; managers on top can be in command of such aspects of their business to a degree they simply could not be before computers.[4]

When it's revealed, near the end of the book, that the computer whose development Kidder has been chronicling is one so large that it has to be disassembled to fit into a freight elevator, the effect, only fifteen years later, is as though one had been reading a history of modern transportation written by someone who'd only driven a Model-T. All the transformations that Kidder ascribes to computer technology are indeed in progress, but the most pervasive transformation—the advent of the personal computer (happening just as Kidder's book appeared)—goes unremarked. The significance of the PC is much like that of PyrE in Alfred Bester's *The Stars My Destination.* PyrE offered the capabilities of nuclear technology to the entire population; the PC democratized computer technology in the same way, and it did so almost overnight. Virtually no one I know has been exempt from the impact of the PC. Computer literacy has become essential to almost any job above the level of busboy. Among those of school age in the '80s, it was the defining difference between being bright or slow, advantaged or disadvantaged. There was probably no better predictor of success among the young than an early fascination with and aptitude for computers.

The same was once said of SF, and at the dawn of the PC age, the two twig-bending influences converged in a new biblio-niche of SF just for hackers—that is, for those who had their own PC and were ready to understand that the distinctive feature of the computer era would not be the emergence of the robot but rather the exploration of a new landscape, cyberspace—not the outer space astronauts were already exploring to diminishing public attention nor the solipsistic Inner Space of the psychedelic '60s. Imagine instead hooking your own neural circuitry directly into the global computer network and entering, like a latter-day Alice stepping through the screen of a computer monitor, fields of surging data.

Here is how William Gibson described the first, toe-wetting stage of a trip into cyberspace:

> Case's virus had bored a window through the library's command ice. He punched himself through and found an infinite blue space ranged with color-coded spheres strung on a tight grid of pale blue neon. In the nonspace of the matrix, the interior of a given data construct possessed unlimited subjective dimension; a child's toy calculator, accessed through Case's Sendai, would have presented limitless gulfs of nothingness hung with a few basic commands.[5]

As to what or where cyberspace is, Gibson finesses the question:

> an abstract representation of the relationships between data systems . . . the electronic consensus-hallucination that facilitates the handling

> and exchange of massive quantities of data . . . mankind's extended electric nervous system, rustling data and credit in the crowded matrix, monochrome nonspace where the only stars are dense concentrations of information, and high above it all burn corporate galaxies and the cold spiral arms of military systems.[6]

If, as I suggested, the automobile is the secret meaning of the rocket ship, then the secret meaning of cyberspace is simply the screen of one's video monitor, a screen that engulfs its user so perfectly that all his sensory data is computer generated. It should come as no surprise that stories featuring virtual reality have become a staple of cyberpunk fiction. The term first appeared in an SF story in 1982 (*The Judas Mandala* by Damien Broderick); by 1995 it commanded an entry of twenty-four column-inches in *The Encyclopedia of Science Fiction*, which cites proto-cyberpunk novels by Philip Dick, Ursula Le Guin, John Varley, and Roger Zelazny, as well as by many official "neuromantics" (as the cyberpunk writers are known collectively, in deference to Gibson's seminal novel, *Neuromancer*).

The problem with virtual reality as an SF concept is that it is so protean as to allow any phantasmagoria to be passed off as science fiction. And that has been just the use to which it was put by the literary avant-gardists and their academic apologists who have tried to join their bandwagon to that of the cyberpunks, a process of co-optation most flagrantly exemplified in the late Kathy Acker's *Empire of the Senseless*, in which the author simply incorporates into her own work those chunks of Gibson's *Neuromancer* that have won her special approbation.

The most zealous academic exponent of the literary claims of cyberpunk SF writers and their avant-garde fellow travelers is Larry McCaffery, editor of *Storming the Reality Studio: A Casebook of Cyberpunk and Postmodern Fiction* (1991). Among the contributors who are accorded honorary cyberpunk status are Kathy Acker, William S. Burroughs, Don DeLillo, Rob Hardin, Thomas Pynchon, William T. Vollman, and Ted Mooney. What their works have in common, to judge by McCaffery's excerpts, would be (1) systematic, lighthearted transgressivity (the "punk" component in cyberpunk) and (2) systematic, lighthearted discontinuity in presenting narrative.

The fracturing of ordinary narrative continuity has been a standard component of dadaism and surrealism for nearly a century. Whether it is done mechanically, as when Burroughs scissors specimens of prose and reassembles the snippets, or the writer relies solely on Errata, the muse of dissociation, the result often can seem to have the jolt of genuine poetry. But only for a while—a paragraph or a page. For any longer stretch, there must be some kind of narrative armature to sustain interest and to be a foil for the surreal ornaments, a semblance of a story, but a really *dumb* story by way of signaling to those readers too hip to read a story that they are actually in the presence of raw transgression. Here is a demonstration of the technique by Mark Leyner, one of McCaffery's candidates for cross-over cyberpunk status:

> I was driving to Las Vegas to tell my sister that I'd had Mother's respirator unplugged. Four bald men in the convertible in front of me were picking the scabs off their sunburnt heads and flicking them onto the road. I had to swerve to avoid riding over one of the oozy crusts of blood and going into an uncontrollable skid. I maneuvred the best I could in my boxy Korean import but my mind was elsewhere. I hadn't eaten for days. I was famished. Suddenly as I reach the crest of a hill, emerging from the fog, there was a bright neon sign flashing on and off that read: FOIE GRAS AND HARICOTS VERTS NEXT EXIT. I checked the guidebook and it said: *Excellent food, malevolent ambiance.* I'd been habitually abusing an illegal growth hormone extracted from the pituitary glands of human corpses and I felt as if I were drowning in excremental filthiness but the prospect of having something good to eat cheered me up.[7]

Leyner continues in the same vein for a while longer, mixing banalities emblematic of American mass culture with moments of off-the-wall grossness illumined by flashes of chic, which are present to assure us that the author can't really be the sophomoric jerk he comes on as. This recipe, with some variations of tempo and inanity, is pretty much the same as that to be found in the selections from Acker, Burroughs, and the other countercultural types represented.

That such work can be credited as a kind of SF reflects the mainstream literary culture's idea that SF equates, simply, with weird and trashy. That is how the French protosurrealist Raymond Roussel, the William Burroughs of his era, responded to Verne's novels. Writers of the New Wave in the '60s were influenced by Roussel, usually indirectly, by their reading of such latter-day Rousselians as Eugene Ionesco, Italo Calvino, and, especially, Harry Matthews, whose aesthetic they often smuggled into their SF. I may have been the chief offender in this regard, at least in the view of *The Encyclopedia of Science Fiction*, where in its article on "Oulipo" (an acronym for *L'Ouvroir de Literature Potenialle*, or "workshop of possible fictions"), I am credited with, in 334, "the most successful Oulipo-related experiment in the SF field." Since I hadn't then read Roussel or heard of Oulipo, I must insist that this was an inadvertent success and that my intent was almost the opposite. Indeed, like Verne, I wanted to write as realistic a novel as I could about the probably near future of the welfare state.

To my mind a "realism of the future" has been the ambition of most good SF writers. The worlds they describe and the events they narrate may have a surreal quality at first glance, but as the story unfolds, such surrealities come to have a naturalistic basis in an altered but *real* world. No such ambition informs the work of Oulipoeans, who take their surrealism neat—a legitimate but very different preference.

The romance between SF and the counterculture has been going on a long time. Since at least the late '60s the spacier elements of the avant-garde have been attracted to SF as a mine of campy pop icons. SF intellectuals, like John Sladek, R. A. Lafferty, Carol Emshwiller, and many more, felt a reciprocal attraction to the disjunctive special effects of high-modernist surrealism and absurdism—usually in short stories that have not made their way into the canon of classic SF, because they were not to the taste of anthologists or, presumably, the fans. Hybrids are delicate organisms.

What chiefly drew avant-gardists to SF was not its aesthetic potential but (as for most readers of whatever brow) its usefulness as a boutique of alternative livestyles, especially if they celebrated sex, drugs, and the right wardrobe. The "punk" component provides exactly the same service for the contemporary youth market, an audience of adolescent and collegiate hackers distinguished from the mere nerds of yesteryear by a larger tolerance for psychotropic drugs and a piercing or two. Here is the hero that Gibson designed for them in the all-time cyberpunk classic of 1984, *Neuromancer:*

> Case was twenty-four. At twenty-two, he'd been a cowboy, a rustler, one of the best in the Sprawl. He'd been trained by the best, by McCoy Pauley and Bobby Quine, legends in the biz. He'd operated on an almost permanent adrenaline high, a byproduct of youth and proficiency, jacked into a custom cyberspace deck that projected his disembodied consciousness into the consensual hallucination that was the matrix. A thief, he'd worked for other, wealthier thieves, employers who provided the exotic software required to penetrate the bright walls of corporate systems, opening windows into rich fields of data.[8]

And here's the heroine, Molly, as we first meet her:

> [Case] realized that the glasses were surgically inset, sealing her sockets. The silver lenses seemed to grow from smooth pale skin above her cheekbones, framed by dark hair cut in a rough shag. . . . She wore tight gloveleather jeans and a bulky black jacket cut from some matte fabric that seemed to absorb light. "If I put this dartgun away, will you be easy, Case?"
>
> . . . The fletcher vanished into the black jacket. . . . She held out her hands, palms up, the white fingers slightly spread, and with a barely audible click, ten double-edged, four-centimeter scalpel blades slid from their housings beneath the burgundy nails.[9]

In the world of cyberpunk, Molly's dominatrix gear has become as comme il faut as pajamas on *Star Trek.* For boys, most kinds of macho grunge will do. And why not? To the degree that SF is just daydreaming, people in novels will wear what the readers' favorite rock stars do. There is, however, an aspect to the "punk" in cyberpunk that can't be reduced to a fashion statement: its acquiescence in the amoral sharkpool politics of the '80s, its acceptance of urban squalor, global pillage, and systemic criminality as the facts of life. Earlier SF writers tended to create classless high-tech utopias or dystopian hells in the spirit of *1984.* The neuromantics didn't buy in to either of these political attitudes. The social vision of cyberpunk is one of nonchalant but profound cynicism. Like the computer hackers who are the real-world counterparts of the "cowboys" and "rustlers" of cyberspace, Gibson & Co. think that crimes committed against MegaCorp and UniGreed are just good sport. Cyberpunk neither glosses over the evils of the real world nor entertains hope of a Le Guinishly sound New Jerusalem. In the cyberpunk future, Third World poverty belongs to everyone, and the American dream has gone belly up. This vision—call it Pop Despair—is ameliorated only by two elements: fashion and an interior life lived in cyberspace. The best hope it offers is that one may be in possession of one's own sensorium, and in that it is a literature designed to reconcile the youth of middle-class America to their lot, to what they've got: sex, drugs, and rock 'n' roll; their hair, their skin, and the clothes on their backs; the fizzing pixels on the video monitor. These are increasingly the options enjoyed and celebrated by Generation X in our own time; tomorrow the fans of cyberpunk will be the managers of the world they are imagining now.

In its essential solipsism, it is also the world that SF has been assiduously constructing since the time of Poe. In both his occluded "philosophical" writings like "Eureka" and in such well-wrought fictions as "The Fall of the House of Usher," Poe hyopthesized worlds that were diagrams of his own psyche, that house and its surrounding dreamscapes being the allegorical externalizations of the poet's disintegrating mind, worlds as entirely (and merely) symbolic as cyberspace

In no single SF writer's work does this tendency bulk so large, or become so explicit, as in that of Robert Heinlein. In the index to H. Bruce Franklin's *Robert Heinlein: America as Science Fiction* there are nine citations under the heading of "solipsism," many of them covering several pages. Nor can this be viewed as the critic's overinterpreting. The notion that the whole world exists only in the imagination of a protagonist much resembling Heinlein is recurrent in his work, as are plot devices that allow all the characters in a book or story to be "cloned" from the Heinlein surrogate. This tendency reaches its apotheosis in the three late novels, *I Will Fear No Evil* (1970), *Time Enough for Love* (1973), and *The Number of the Beast* (1979). Their plots are a tangle of narcissistic love knots, which, when untied, come down to these words of advice from *Time*

Enough for Love, given by God to Lazarus ("Call Me Heinlein") Long, when he'd expressed a desire to see God's face: "Try a Mirror."

"His maker is himself," Franklin comments, "and . . . Lazarus is trapped in a solipsistic world of his own devising, one where all other beings are merely reflections of himself." That comment is literally the synopsis of Heinlein's comic apocalypse, *The Number of the Beast*, which culminates in a parody of an SF convention entitled "The First Centennial Convention of the Interuniversal Society for Eschatological Pantheistic Multiple-Ego Solipsism," where all Heinlein's favorite characters from his own books and those he's made his own by having read them have gathered from the various universes where all fictions, but Heinlein's especially, have been made flesh. The purpose of the gathering is to celebrate Heinlein's divinity. The last chapter is entitled "Rev. XXII:13," the verse of the book of Revelations: "I am the Alpha and Omega, the beginning and the end, the first and the last."

Even with many grains of salt, one must wonder how Heinlein expected readers to view this revelation. Surely not as the gospel truth, even if interpreted with New Age liberality as meaning that each man is his godhead—but not as an ironic jape, either. Rather, it is the freakout to which he's entitled as a good American, whose right to lie is protected by the Constitution.

If Americans do believe they have a right to lie, as I maintained in Chapter 1, then its philosophical basis must be the radical solipsism that SF has always allowed as a fundamental premise: I am the Alpha and Omega; I've been abducted by aliens; the speed of light can be exceeded; I hunt for dinosaurs in my time machine every other Thursday; I may be fat but I'm a telepath, so beware. Anything goes, if it's a satisfying daydream.

Heinlein is hardly the only author to have made that discovery. In 1979 John Varley, then the most celebrated new writer in the field, published "The Persistence of Vision," a story in which a group of disastrously disabled (but psychically gifted) New Agers achieve nirvana and semiomnipotence by virtue of the secret wisdom only paraplegics possess—a vindication of the fannish wisdom that "reality is a crutch." Predictably, SF fans loved it, and it won both the Hugo and Nebula awards.

The moral imperative of Varley's story is *Wish really hard.* That of Larry Niven's novel *Ringworld* (1970), which also swept the awards, is *Get lucky.* The denouement reveals that his protagonists, genetically selected for luck, can only come to good. So much for narrative suspense!

A more highbrow version of Niven's injunction appears in Greg Egan's first novel, *Quarantine* (1992). Egan, a young Australian, has been to the '90s what Varley was to the '70s: the most likely to succeed. He does wonderful extrapolations of nanotechnology—the possibility of creating microscopic machineries with viral and cybernetic components; in effect, autonomous, intelligent drugs—but he doesn't stop there. Instead of creating a *human* drama around that premise, he uses quantum theory, that perennial excuse for Anything Goes, to power a plot that, once more, results in another Alpha-and-Omega protagonist, who is literally unable to come to a bad end because bad ends can come only to his infinitely many clones in the infinitely many universes secondary to this novel, in which he's *logically* guaranteed a happy ending. As with Heinlein and SF's other prime solipsists, the lesson to be drawn from Egan is, "I think, therefore I am the creator God." Egan says this more cleverly than his predecessors, and his arabesques are often hypnotically intricate, but just as the Koran proscribes representations of the human figure, solipsism proscribes human drama. When you can only win, why tell tales?

In the near future—the next ten years, say—inertia should keep the world of SF looking much as it has been described here. Already even the most artful and ambitious of SF novelists have adapted their talents to the exigencies of the marketplace and are producing not singletons but product. Gene Wolfe, Kim Stanley Robinson, William Gibson: each completes one trilogy, only to begin another. Novelists who do not make such an adaptation will find themselves squeezed to the sidelines of university and small press publication, like poets and nongenre midlist novelists. Necessarily, if they are writing chiefly for love or academic credentials, they will write less abundantly, and many will give a larger part of their writing time to academic criticism, since that is likelier to advance their careers in academia.

This has been the standard c.v. for many of the most notable writers of the New Wave era. Only a few SF writers of an earlier era (Edmund Hamilton and James Gunn notably) had the opportunity or took the trouble to become credentialed in academia, but since the '70s, a significant number of the top-rank SF writers have pursued double careers. Gregory Benford is a physicist at the University of California, San Diego; Joanna Russ teaches at the University of Washington, Joe Haldeman at MIT, John Kessel at North Carolina State University, Scott Bradfield at the University of Connecticut. The list goes on.

No other SF writer has had a more representative and instructive career in this regard than Samuel R. Delany. One of the genre's many teen prodigies, his first novel appeared in 1960, when he was twenty. In 1966, after a succession of apprentice paperback novels for Ace Books, Delany brought out his first distinctively New Wave work, *Babel 17*, which was followed by *The Einstein Intersection* (1967), *Nova* (1968), and his SF magnum opus, *Dhalgren* (1975). He garnered many awards for these books and for his short fiction, but then, having clearly established his preeminence as the American New Wave's most brightly shining star (having been most conspicuously ambitious and successful without forfeiting an aura of genre funk), Delany shifted gears. His fiction production slowed down in the '80s and became, especially in his *Neveryon* phantasmagorias, ever more tendentiously the podium for his non-SF interests: the intellectually intertwined realms of deconstructive literary criticism and queer theory. At his nadir, he produced a novel/memoir/diatribe, *The Mad Man*, with the doubtful thesis that HIV is *not* the cause of AIDS, a favorite lost cause among queer theorists, the late Michel Foucault among them.

In summarizing Delany's meteoric career, Peter Nicholls writes, in *The Encyclopedia of Science Fiction*, "With hindsight it can be hypothesized that Delaney has had different audiences at different points of his career: a very wide traditional SF readership up to and including *Dhalgren* . . . and a narrower, perhaps more intellectual, campus-based readership thereafter."[10] Ursula Le Guin and Joanna Russ have shared Delany's fate, to some degree, but since their (feminist) sexual politics are less radical than Delany's (whose three pornographic novels are as doctrinairely transgressive as de Sade's), they have maintained an emeritus standing within the genre, while Delany seems to have changed his permanent address from science fiction to academe.

As to the future of SF, apart from the fortified suburbs of tenured teaching, the outlook is bleak. Overall genre sales are in decline, lists are shrinking, and media tie-ins continue to dominate paperback sales—so long as the series that inspire them remain popular, at least in syndication. However, all media are mortal, and *Star Trek* is once again rumored to be slated for termination. A further sign that the franchises are in trouble is word from the editorial grapevine that those who generate franchise product will soon be writing for hire only, sans even a 2 percent slice of the pie. I would infer from this that the franchise owners foresee a near future in which megabuck epics and TV space operas will soon be as rare as the once-ubiquitous horse operas. Such has ever been the fate of cash cows and golden-egg-laying geese.

Finally, one must consider the fate of the book itself. So far the early alarums about the computer's supplanting the printed page have been unfounded. The printed page is still cheaper and more convenient than prose encoded on disc. But once that technology is in place, SF is likely to be the first genre to cross the technological gap in a big way. SF fans, after all, were among the first enthusiasts of the personal computer, and a good percentage of them are amateur writers who would welcome the opportunity—already provided by the more prescient SF authors, like Orson Card—of kibbitzing rather than passively receiving the writer's word as writ. Interactive fiction and hypertext are still the pastimes of a small minority, but in ten or twenty years, with the equipment in place, they might well supplant a good part of ordinary genre fiction. The deductive mystery is a natural candidate for such a transformation, and SF stories of first contact and of exploring alien planets or visiting utopias would seem to lend themselves to an interactive slant. But that is still a long way down the road.

From my own experience writing an interactive novel, I can safely predict that once such works become common, they will be created, like TV shows, by teams of writers rather than by single authors. The amount of prose that must be written to flesh out a large fictional environment (rather than to tell a linear story) will doom the soloist author to extinction. As for soap operas and open-ended dramatic series, an orchestra of writers will be required. We viewers will not know their names. They'll scroll by too quickly.

What those who enjoy fiction of whatever sort primarily seek is vicarious involvement, and they would rather *see* a tale being told than read it. *Hamlet* in a reasonably good stage production is more engaging than *Hamlet* on the page. SF has enjoyed its edge among readers for so long because until very recently it could tell stories that could not be filmed. Now it is budgets, and not technology, that dictate what is possible. The same computers that have made publishing so much easier and cheaper have done the same for visual media.

Movies and TV are bound to win. Huxley knew that when he described a night at the "feelies" in *Brave New World,* but Huxley assumed that such a transformation would lead to a universal leveling to the lowest common denominator, and we would all be wallowing in porn every Saturday night. In fact, the index of good taste has risen even as it has sunk. The menu of prime-time TV includes operas and horror movies, professional wrestling and intense realizations of the novels of Jane Austen, Charles Dickens, and any other novelist who may take a producer's fancy.

As to interactivity, which is theoretically the ideal goal of the possible relationship of an author and his gentle readers, Ray Bradbury, almost half a century ago, foresaw its technical inevitability. At the dawning of the age of television he wrote *Fahrenheit 451*, in which the anomic wife of the protagonist (a "fireman" devoted to burning all books and the houses that might contain them) is a devotee of interactive soap opera. Here's her account of what the genre offers:

> "This is a play that comes on the wall-to-wall circuit in ten minutes. They mailed me my part this morning. I sent in some boxtops. They write the script with one part missing. It's a new idea. The homemaker, that's me, is the missing part. . . . Here, for instance, the man says, 'What do you think of this whole idea, Helen?'"[11]

What Bradbury thinks of the whole idea is obvious: the death of Western culture.

In terms of the technological possibilities of the age of television,

Bradbury was prescient, but in terms of the cultural impact of the media, he was dead wrong. People now have more information, and they are smarter, overall, as a consequence—even in those ways they choose to be dumb.

Delmore Schwartz had half of it right: in dreams begin responsibilities. But it's no less true that in dreams begin irresponsibilities. The menu, in terms of our possibilities in both those respects, is well-nigh infinite.

Science fiction is that menu.

NOTES AND REFERENCES

Introduction

1. Kenneth Fearing. "No Credit," in *Collected Poems of Kenneth Fearing* (New York: Random House, 1940), p. 59.

2. Lester del Rey. "Helen O'Loy," *Modern Science Fiction.* Norman Spinrad, editor. (Garden City, NY: Anchor, 1974), pp. 58, 67.

Chapter 1

1. Miller himself may have accounted Abigail a villain, but since the first appearance of the play in 1953, reports of ritual satanic child abuse have become so widespread, and so widely believed, that Abigail must now be accounted a kind of role model for those of a similar, neo-Puritan bent.

2. Sissela Bok. *Lying: Moral Choice in Public and Private Life* (New York: Vintage, 1979), p. 149.

3. Phillip Knightley. *The First Casualty: From the Crimea to Vietnam: The War Correspondent as Hero, Propagandist, and Myth Maker* (New York: Harcourt Brace Jovanovich, 1975), p. 376.

4. Ibid., p. 423.

5. Ben Bradlee, Jr., *Guts and Glory: The Rise and Fall of Oliver North* (New York: Donald I. Fine, 1988), p. 544.

6. The classic example is *Secret Survivors* by E. Sue Blume (New York: Wiley, 1990). Among the thirty-four categories of symptoms of repressed incestuous abuse on Blume's "Incest Survivors' Aftereffects Checklist" are: poor body

image, headaches, arthritis, drug or alcohol abuse (or total abstinence), "having dreams or memories," stealing, phobias, and a "need to be invisible, perfect, or perfectly bad." It is hard to imagine anyone, male or female, who could exempt themselves from suspicion of being an incest survivor, given the checklist's damned-if-you-do, damned-if-you-don't comprehensiveness. But that, of course, is the book's purpose. It is an invitation—to quote the blurb by Elizabeth Kübler-Ross on the back cover of the paperback edition—"for those who suspect that they are unconscious survivors of abuse and especially for therapists to dig into the darkest shadow part of human existence."

7. Whitley Strieber, *Communion: A True Story* (New York: Morrow, 1987) and *Transformation: The Breakthrough* (New York: Morrow, 1988).

8. Their 1961 adventure was set down by John G. Fuller in *The Interrupted Journey* (New York: Dial, 1966), but had its full impact only in 1975, when NBC broadcast a two-hour version, *The UFO Incident*. Now such pseudodocumentaries, recreating the whopper of the week, are a regular feature of prime time on such series as *The Paranormal Borderline*.

9. Douglas Curran, *In Advance of the Landing: Folk Concepts of Outer Space* (New York: Abbeville, 1985), p. 21. Curran's book is remarkable for the author's gallery of photographs of homemade flying saucers that decorate the American landscape, as well as for Curran's sympathetic but closely observed accounts of the lifestyles of the most true-believing UFO enthusiasts.

10. Strieber, *Communion: A True Story*, p. 243.

11. Bernice Kanner, "Americans Lie, or So They Say," *New York Times*, May 30, 1996.

12. Benford was allowed to express a more candid estimation of the book when he was interviewed for an article in *Publishers Weekly*, "When Is a True Story True?" (August 14, 1987): "This book is part of a deplorable trend in publishing. It is catering to the flagrant irrationalities of the public with tarted-up Potemkin-Village science. The re-emergence of the Shirley MacLaine/Bridey Murphy subgenre is a chastening reminder that we are not, in fact, a deeply rational society in spite of our technology."

13. Thomas M. Disch, "The Village Alien," *Nation*, March 14, 1987.

14. Philip J. Klass, *UFO Abductions: A Dangerous Game* (Buffalo: Prometheus, 1989), p. 208.

15. Ignatius Donnelly, *Atlantis: The Antediluvian World* (New York: Dover, 1976), p. 1.

16. Erich von Daniken, *Chariots of the Gods* (New York: Bantam, 1971), pp. 51–52.

17. Jasper Griffin, "Anxieties of Influence," *New York Review of Books*, June 20, 1996, p. 70.

18. James Wolcott, "I Lost It in the Saucer," *New Yorker*, July 31, 1995, p. 77.

Chapter 2

1. J's claim was appreciably strengthened in 1990, when Ted Chiang's novelette, "Tower of Babylon," won a Nebula Award and was nominated for a Hugo. Chiang's tale is a Babylonian technothriller recounting what might have happened if another Nimrod had succeeded in building a tower that reached heaven. The story is scientifically accurate in all its details, bearing in mind that its informing cosmology is that of ancient Sumeria.

2. Brian W. Aldiss. *Billion Year Spree: The True History of Science Fiction* (New York: Schocken, 1974), p. 26.

3. Jeffrey Meyers, *Edgar Allan Poe: His Life and Legacy* (New York: Scribner's, 1992), p. 261.

4. Jeffrey Meyers, quoting from Whitman's *Specimen Days. Edgar Allan Poe: His LIfe and Legacy* (New York: Scribner's, 1992), p. 265.

5. T. S. Eliot, "From Poe to Valery" (1948), in *To Criticize the Critic* (New York: Octagon, 1965), p. 35.

6. Poe's zeal to become an editor is another mark of his prescience. Since *Amazing Stories* was established in 1926 by Hugo Gernsbach (after whom the field's oldest award, the Hugo, is named), it has been the magazine editors who have set the genre's agenda: John Campbell (*Astounding*, which became *Analog*), Frederik Pohl (*Galaxy*, *If*), Michael Moorcock (*New Worlds*), Gardner Dozois (*Asimov's*). The machine, in SF, is more important than the individual man.

7. Daniel Hoffman, *Poe, Poe, Poe, Poe, Poe, Poe, Poe* (Garden City, N.Y.: Doubleday, 1972), p. 159.

8. "Einstein's brain is a mythical object: paradoxically, the greatest intelligence of all provides an image of the most up-to-date machine. . . . The mythology of Einstein shows him as a genius so lacking in magic that one speaks about his thought as of a functional labor analogous to the mechanical making of sausages." Roland Barthes, *Mythologies* (New York: Hill & Wang, 1957).

9. The history of SF recapitulates the history of literature in this respect, as in many others. In the '60s and '70s the large menu of psychotropic drugs illuminated the work of an entire generation of SF writers (the generation I belong to)—with similar after-the-fireworks results: fizzle and burnout. More on that anon.

10. Jeffrey Meyers. *Edgar Allan Poe: His Life and Legacy.* (New York: Scribner's, 1992). p. 180.

11. Timothy Spencer Carr, "Son of Originator of 'Alien Autopsy' Story Casts Doubt on Father's Credibility," *Skeptical Inquirer* (July–August 1997).

12. Peter Washington, *Madame Blavatsky's Baboon* (New York: Schocken, 1993, 1995), p. 36.

13. Quoted from John Symonds, *Madame Blavatsky—Medium and Magician* (London: Odhams, 1959).

14. Whitley Strieber, *Communion: A True Story.* (New York: Morrow, 1987).

Chapter 3

1. The publication of of *De la terre à la lune* marks the beginning of the 101-year flight time of this chapter's title. The landing of the Soviet *Luna* 9 in 1966 marks its close.

2. But not without a preliminary salute to our genre's founding genius. "Cheers for Edgar Poe," choruses the Gun Club, in Chapter 2, as they endorse their President's proposal to shoot the first Americans to the moon.

3. Also . . . or perhaps for that very reason. It can be argued that any effective novel is also a good polemic, though its agenda may often be obscured by its verisimilar effectiveness. Wells had a memorable set-to on just this point with Henry James, who after applauding Wells's first novels vigorously, began to find them tendentious. As they were. It is harder to discern James's own tendency at times, but it is there. No man is without an agenda.

4. The figure of the dinosaur is, in that respect, one of the most potent "mythological" figures in the SF repertory: *The Beast from 20,000 Fathoms* (1953), based on a Ray Bradbury story, *Godzilla* (1954, and a remake in production as I write), and *Jurassic Park* (1994). The dinsosaur is the modern age's most popular monster.

5. Richard J. Herrnstein & Charles Murray. *The Bell Curve: Intelligence and Class Structure in America* (New York: Free Press, 1996).

6. Or, more risibly, the mammalian equivalent of the bluefooted booby, a bird whose character is expressed by its name. Such is the thesis of Kurt Vonnegut's last great work of science fiction, *Galapagos* (1985), in which the passengers on a cruise ship, the *Bahia de Darwin*, are shipwrecked in the Galapagos Islands just as the rest of the human race self-destructs. The new homo galapagos endures a million years precisely because they devolve into the condition of seals and walruses, free of their ancestors' *excessive* intelligence and imagination.

7. In his persistent sequelizing of any significant success, Verne prefigures another SF tradition: milk any cash cow dry.

8. Arthur C. Clarke, *2001: A Space Odyssey* (New York: Signet, 1968), p. 218.

9. Hal Lindsey, *The Rapture* (New York: Bantam, 1983), pp. 44–45.

10. More secular authors have posited more secular corollaries of this equation. In *The Void Captain's Tale* (New York: Timescape, 1983), Norman Spinrad hypothesizes, in a Wilhelm Reichian way, that the key to faster-than-light travel is orgasm.

11. Quoted from Paul Boyer's definitive study, *When Time Shall Be No More: Prophecy Belief in Modern American Culture* (Cambridge, Mass.: Harvard University Press, 1992), pp. 7–8.

12. Quoted from Charles Platt, *Dream Makers* (New York: Berkley, 1980), p. 180.

13. Arthur C. Clarke, *Profiles of the Future* (New York: Holt, Rinehart and Winston, 1984), p. 95.

14. David Hartwell, *Age of Wonders* (New York: Walker, 1984), pp. 80–81.

15. J. G. Ballard, "The Message from Mars," in David Pringle (ed.), *The Best of Interzone* (New York: St. Martin's Press, 1997), p. 35.

Chapter 4

1. Ray Bradbury. "There Will Come Soft Rains," in *The Martian Chronicles* (New York: Avon, 1997) p. 249.

2. Twenty-eight years later, Bradbury will be plugging the space program on basically the same grounds: "Because we *can* escape, we *can* escape, and escape is very important, very tonic, for the human spirit." But the "escape" Bradbury is talking about in 1978 is simple escapism, a kind of self-bamboozlement. He explains it very clearly: "From the time I saw my first space covers on *Science and Invention*, or *Wonder Stories*, when I was eight or nine years old—that stuff is still in me. Carl Sagan, a friend of mine, *he*'s a 'romantic,' he loves Edgar Rice Burroughs—I know, he's *told* me." Quoted from an interview in Charles Platt, *Dream Maker* (New York: Berkley, 1980), p. 180.

3. Quoted in Paul Boyer, *By the Bomb's Early Light: American Thought and Culture at the Dawn of the Atomic Age* (New York: Pantheon, 1985), p. 304.

4. Ibid., p. 298.

5. Ibid., p. 305.

6. Ray Bradbury, *The Stories of Ray Bradbury* (New York: Knopf, 1980), pp. 15–16.

7. Postapocalyptic bikers are not true mutants, of course. Properly, they descend from a later, tamer form of doomsday movie that began with *Panic in the Year Zero* (1962), in which nuclear war is the stimulus for survivalist fantasies. In *Panic*, Ray Milland is a dad who shepherds his nuclear family through the anarchy

that follows the bombing of Los Angeles. This time the monster unleashed by the bomb is a trio of elderly juvenile delinquents who first appear in a hot rod. In *The Day the Earth Caught Fire,* a British nuclear catastrophe movie of the same year, anarchy takes the form of a congo line of rebel youth wearing only swimsuits and playing jazz on their saxophones.

8. Ray Bradbury's *The Martian Chronicles* (1950), Bernard Wolfe's *Limbo* (1950), Leigh Brackett's *The Long Tomorrow* (1955), Alfred Bester's *The Stars My Destination* (1956), Philip K. Dick's *Time Out of Joint* (1959), Pat Frank's *Alas, Babylon* (1959), and Walter M. Miller's *A Canticle for Liebowitz* (1959).

9. George Orwell's *1984* (1949), George Stewart's *Earth Abides* (1949), Robert Heinlein's *The Puppet Masters* (1951), John Wyndham's *The Day of the Triffids* (1951), Arthur C. Clarke's *Childhood's End* (1953), Clifford Simak's *Ring Around the Sun* (1953), Edgar Pangborn's *A Mirror for Observers* (1954), and John Christopher's *The Death of Grass* (1956).

10. H. Bruce Franklin. *Robert A. Heinlein: America as Science Fiction* (New York: Oxford University Press, 1980), p. 3.

11. In hindsight, Marcuse's "repressive desublimation" seems simply a last-ditch effort to reconcile late Marxist political theory with Freudian doctrine. In the '60s, sexual mores were loosening up, while capitalism exhibited no parallel progressive tendency, to the distress of ideologues. The disjunction needed a theory and a name; Marcuse supplied both.

12. Robert Heinlein, *The Puppet Masters* (New York: Signet, 1951), pp. 174–175.

13. Ibid., p. 40.

14. Ibid., p. 108.

15. Boyer, *By the Bomb's Early Light*, pp. 341–342.

16. Quoted in Gregg Rickman, *To the High Castle: Philip K. Dick: A Life 1928–1962* (Long Beach: Valentine Press, 1989), p. 119.

17. Ibid., pp. 272–273.

18. The object of a search in a thriller, like *The Maltese Falcon.* Hitchcock always has a McGuffin

19. Alfred Bester, *The Stars My Destination* (New York: Signet, 1957), chap. 16.

20. As his star rose and he began to receive critical attention, Dick got edgy. In a 1976 hardcover reprint of his first novel, *Solar Lottery,* I had written: "*Solar Lottery,* along with most of its successors, may be read as self-consistent social allegories of a more-or-less Marxist bent." Dick, interviewed by his biographer, Gregg Rickmann, in 1978, transformed this observation into an assertion that I'd labeled him "the only Marxist science fiction writer there is," and went on

to take an oath of allegiance. He had earlier, on reading my novel *Camp Concentration,* written to the FBI to report me as a danger to the republic. Envy? Perhaps, but for a true paranoid, such a denunciation may be the sincerest form of flattery. I choose to be flattered.

21. Philip K. Dick, *The Penultimate Truth* (New York: Bluejay, 1984), p. 58.

22. Boyer, *By the Bomb's Early Light,* pp. 355–356.

23. Ibid., p. 358.

Chapter 5

1. Peter Nicholls and John Clute (eds.),*The Encyclopedia of Science Fiction* (New York: St. Martin's Press, 1993), p. 1208.

2. Ibid., p. 1157.

3. *And* the passion can be consummated for a modest price. For $49.95 a dedicated Trekkie can purchase a Star Trek: Deep Space Nine shirt complete with dickie, metallic pips, and emblem. For $59.95 ladies can get a reproduction of Kira's jumpsuit or the bolder Next Generation model. There are uniforms in children's sizes, wristwatches and alarm clocks, "Bajoran" earrings, bracelets, lapel pins, telephones, and key rings, not to mention games, audiotapes, posters, books, and bumper stickers. All available—and more!—from 800-Trekker, the "24 Hour Sci-Fi Collectibles Hotline."

4. Richard Raben and Hiyaguha Cohen, *Boldly Live As You've Never Lived Before* (New York: Morrow, 1995), pp. 212–213.

5. Ibid., pp. 165–166.

6. Ibid., pp. 233–234.

7. Edward Bellamy, *Looking Backward* (New York: Signet, 1960), p. 143.

8. Ibid., pp. x–xi.

9. J. G. Ballard, *The Atrocity Exhibition,* rev. ed. (San Francisco: Re-Search Publications, 1990), p. 9.

10. Michael Moorcock. *The Final Programme* (New York: Avon, 1968) p. 38.

11. A famous criminal, the original "Bluebeard," and friend of Joan of Arc. Also spelled de Rais.

12. Burroughs did enact a pro forma breast-beating with regard to his drug of preference, heroin, which he deplored even as he drooled over it. Other illicit drugs, however, get a green light. In his Introduction to *Naked Lunch* (New York: Grove, 1959), he is explicit: "When I speak of drug addiction I do not refer to keif, marijuana or any preparation of hashish, mescaline, *Bassnisteria Caapi,* LSD6, Sacred Mushrooms or any other drug of the hallucinogen group. . . . There is no evidence that the use of any hallucinogen results in physical depen-

dence. The action of these drugs is physiologically in opposition to the action of junk. A lamentable confusion between the two classes of drugs has arisen owing to the zeal of the U.S. and other narcotic departments." That shopping list of drugs that a wise consumer might safely play with has remained the received wisdom of the counterculture to this day.

13. Aldous Huxley. *The Doors of Perception* (New York: Perennial Library, 1990), p. 18.

Chapter 6

1. U.S. district judge Frederick vanPelt Bryan, quoted in an appendix to the 1959 Grove Press paperback edition of D. H. Lawrence's *Lady Chatterley's Lover.*

2. John Norman, *Rogue of Gor* (New York: DAW, 1981), pp. 108–110.

3. It was this element of communes featuring dorm-style promiscuity under the auspices of bullying alpha male that especially endeared the book to the counterculture of the '60s, including Charlie Manson, who made *Stranger* required reading for his followers.

4. Robert Heinlein, *Stranger in a Strange Land* (New York: Ace, 1991), pp. 441–443.

5. Robert Heinlein, *Friday* (New York: Holt, Rinehart and Winston, 1982), p. 10.

6. Vonda McIntyre, *Fireflood* (Boston: Houghton Mifflin, 1979), p. 87.

7. John Clute, *Science Fiction: The Illustrated Encyclopedia* (New York: Dorling Kindersley, 1995), p. 177.

8. The Hugo Awards reflect the judgment of SF fandom. The Nebula Awards are given out by the Science Fiction Writers of America.

9. Ursula Le Guin, *The Word for World Is Forest* (New York: Berkley, 1976), p. 81.

10. Ursula Le Guin, "American SF and the Other," in *The Language of the Night: Essays on Fantasy and Science Fiction* (New York, Berkley, 1982), p. 87–89.

11. Marleen S. Barr, "*Searoad Chronicles of Klatsand* as a Pathway toward New Directions in Feminist ScienceFiction: Or, Who's Afraid of Connecting Ursula Le Guin to Virginia Woolf?" *Foundation: The Review of Science Fiction* 60 (spring 1994).

12. George Slusser, "The Politically Correct Book of Science Fiction: Le Guin's Norton Anthology," *Foundation: The Review of Science Fiction* 6 (spring 1994). Slusser's essay follows immediately after that by Barr cited in note 11. The ideological distance between the two authors is representative of the untemporizing (if often ill-tempered) liveliness of the U.K.-based *Foundation,*

which has been the most reliable venue for criticism of SF for the past twenty-five years.

13. Slusser's response to this, in ibid., is to ask "why there aren't more white Kabuki dancers, or haiku poets, or Fante flag makers." He also points out "that there *are* Asian and Latin SF writers. The hitch is that they are not Asian-*American* or Latin-*American* writers."

14. Ursula Le Guin, Introduction to *The Norton Book of Science Fiction: North American Science Fiction, 1960–1990* (New York: Norton, 1993), p. 17.

15. Barry N. Malzberg, "Making It All the Way into the Future on Gaxton Falls of the Red Planet," in *Norton Book,* p. 313.

16. It is a recipe worth passing on. While it may have too much of eggs and sugar to be politically correct among the stricter sort of dietitians (I fear that Ursula never tried it out), let me commend it again here:
Blend: 1 cup sugar; 5 tbsp flour; 2 tbsp butter; pinch of salt
Add: 1/4 cup lemon juice; zested rind of ½ lemon; 3 egg yolks (beaten), 1 cup milk
Mix: Fold in 3 beaten egg whites.
Pour in a buttered baking dish, set in pan of hot water and bake 45 minutes in a 350-degree oven. Don't overbake. The top part gets cakey but the bottom stays a little slurpy, so that it makes its own sauce.

17. Joanna Russ. "When It Changed" in *Again, Dangerous Visions,* edited by Harlan Ellison (New York: Berkley, 1983), p. 254.

18. Robert Heinlein, *I Will Fear No Evil* (New York: Putnam, 1970).

19. Jessica Amanda Salmonson, "Our Amazon Heritage," in *Amazons!* (New York: DAW, 1979).

Chapter 7

1. David E. Kaplan and Andrew Marshall, *The Cult at the End of the World* (New York: Crown, 1996), p. 31.

2. Willa Cather and Georgine Milmine, *The Life of Mary Baker G. Eddy and the History of Christian Science* (Lincoln: University of Nebraska Press, 1993). Milmine appeared as the sole author on the first publication of the book in 1908, as a serial in *McClure's* magazine, but her contribution was the research; Cather was the actual author.

3. *The Encyclopedia of Science Fiction* (New York: St. Martin's, 1993).

4. Campbell's track record for discriminating between science and hoax was not good. After parting ways with Hubbard, he championed the Heironymous machine in 1956, a "psionics" machine akin to a Ouija board that, wonderfully,

works as well as a diagram on paper as when built from metal and plastic—a Platonic idea incarnate. In 1961, Campbell endorsed the Dean Drive, an antigravity device in the immemorial tradition of perpetual motion machines. Campbell, for all his insistence on "hard" science, was always ready to ballyhoo any will-o'-the-wisp pseudoscience that he knew would boost his circulation for a month or two.

5. John W. Campbell. Introduction to *Dianetics: The Modern Science of Mental Health* (New York: Hermitage House, 1950).

6. Russell Miller, *Bare-Faced Messiah: The True Story of L. Ron Hubbard* (New York: Henry Holt, 1987), p. 157.

7. Ibid., p. 156.

8. Ibid., p. 203.

9. Quoted from Lawrence Sutin, *Divine Invasions: A Life of Philip K. Dick* (New York: Harmony, 1989), p. 211.

10. Ibid., p. 213.

11. Ibid., p. 219.

12. Ibid., p. 227. From an interview in June 1986, quoted with some mistranscriptions emended.

13. Ibid.

14. Temporal lobe epilepsy (TLE) has also been offered as an explanation for the reported experiences of Whitley Strieber and other UFO abductees. In Strieber's case I have my doubts, but in some of the earlier and more naive cases, TLE may have played a role.

15. Sutin, *Divine Invasions*, p. 231.

16. Ibid.

17. Phillip Dick. "The Exegesis" (excerpts), in *Gnosis* (fall–winter 1985).

18. TK.

19. Ibid., p. 10.

20. Joseph Campbell, *The Hero with a Thousand Faces* (Cleveland: Meridian, 1963), p. 230.

21. Kinney, ((()), p. 9.

22. Gerald Jonas. "The Shaker Revival," in Thomas M. Disch (ed.), *The Ruins of Earth* (London: Hutchinson, 1973).

Chapter 8

1. Quoted from *PKD: A Philip K. Dick Bibliography*, comp. Daniel J. H. Levack (San Francisco: Underwood/Miller, 1981), p. 116.

2. The sleuthing behind this story was the work of Thomas Perry. The entire

account of Heinlein's efforts to cover his EPIC footprints may be found in Perry's "Ham and Eggs and Heinlein," which appeared in 1993 in the third number of Damon Knight's brief-lived magazine, *Monad.*

3. Robert A. Heinlein, *Starship Troopers* (New York: Berkley, 1968), p. 147.

4. I. F. Clarke, *Voices Prophesying War,* new ed. (New York: Oxford University Press, 1992), p. 29.

5. Ibid., p. 41.

6. Richard Hofstadter. *The Paranoid Style in American Politics: And Other Essays* (Chicago: U. of Chicago Press, 1979).

7. Ignatius Donnelly (writing as Edmund Boisgilbert, M.D.), *Caesar's Column: A Story of the Twentieth Century* (Chicago: F. J. Schulte & Co., 1891).

8. Ibid., p. 113.

9. Jerry Pournelle, quoted in *Dream Makers, Vol. 3: Interviews by Charles Platt* (New York: Berkley, 1983), p. 7.

10. Ibid., p. 8.

11. Ben Bova, *The High Road* (New York: Pocket Books, 1983), p. 207.

12. Jerry Pournelle and Dean Ing, *Mutual Assured Survival* (Riverdale, N.Y.: Baen, 1984).

13. Baen Books blurbs Drake's 1991 opus, *The Warrior,* in these glowing terms: "They were the best. Colonel Alois Hammer welded five thousand individual killers into a weapon more deadly than any other in the human universe. But different styles of being 'the best' meant a bloodbath, even by the grim standards of *Hammer's Slammers.*"

14. Garry Wills, "It's His Party," *New York Times Magazine,* August 11, 1996.

15. *New York Times,* March 30, 1986.

16. *Times Herald Record* (Middletown, N.Y.), September 11, 1996.

17. Kristin Hunter Lattany, "Off-Timing: Stepping to the Different Drummer," in Gerald Early (ed.), *Lure and Loathing* (New York: Penguin, 1994).

18. Warren Leary, "Space Agency Plans Layoffs, Shrinking to Pre-Apollo Size," *New York Times,* May 20, 1995, p. 1.

19. Yet another collaborator on *Window of Opportunity,* though credited only among the acknowledgments, was Janet Morris, an SF writer and a consultant in weapons development. She was one of the co-drafters, along with C. J. Cherryh (who specializes, like Morris, in tales of future warfare), of the March 30, 1986, "Letter to America" in the *New York Times.*

20. Newt Gingrich, *To Renew America* (New York: HarperCollins, 1995), p. 192.

21. Ibid., p. 191.

22. Ibid., p. 189. One must wonder what books of H. G. Wells Gingrich has in

mind. His first novel, *The Time Machine*, with its devolved class system of effete Eloi and cannibalistic Morlocks? *The Island of Dr. Moreau?* His deathbed confession of despair for the fate of the species, *A Mind at the End of Its Tether?* His assessment of Crichton's adventure story as pessimistic is similarly off the wall.

23. This series was originated by Ed Naha, but most of the later titles, according to the *Encyclopedia of Science Fiction* (New York: St. Martin's Press, 1993), have been the pseudonymous work of the cyberpunk writer John Shirley.

24. *Encyclopedia of Science Fiction*, p. 646.

25. Robert Heinlein, *Farnham's Freehold* (Riverdale, N.Y.: Baen, 1994), p. 33.

26. The SF writers who have tried their hand at historical fiction (not counting those who have written hybrids of the two genres that involve traveling backward in time or "alternative" pasts) include Jules Verne, John Brunner, Ursula Le Guin, Robert Silverberg, Brian Aldiss, Orson Scott Card, and me. The reason for this crossover phenomenon lies in the similarity of the task: to create a densely imagined world, with social protocols and physical environments radically unfamiliar to most readers. That skill, learned in one genre, can be readily transferred to the other.

27. Quoted by Neil Gaiman and Kim Newman (eds.), *Ghastly Beyond Belief* (New York: Arrow, 1985), p. 78.

Chapter 9

1. Emanuel Swedenborg, *De telluribus* (1758), quoted from the 1787 English translation.

2. John Adams, "Outer Space and the New World in the Imagination of Eighteenth-Century Europeans," *Eighteenth Century Life* (February 1995).

3. H. G. Wells. *The War of the Worlds* (New York: Popular Library, 1962), p. 131.

4. Ibid., p. 163.

5. A. E. Housman. "Epitaph on an Army of Mercenaries," reprinted as the epigraph to William Barton's *When Heaven Fell* (New York: Warner, 1995) front matter.

6. *Times-Herald Record* (Middletown, N.Y.), August 16, 1996, p. 50.

7. James Tiptree, Jr., "The Women Men Don't See," quoted in *The Norton Book of Science Fiction* (New York: Norton, 1993), p. 271.

8. James Bowman, "American Notes," *London Times Literary Supplement*, August 9, 1996.

9. The Alvin Maker novels are, in order of publication, *Seventh Son* (1987), *Red Prophet* (1988), *Prentice Alvin* (1989), and *Alvin Journeyman* (1995), with more volumes projected.

10. Orson Scott Card, *Prentice Alive: The Tales of Alvin Maker III* (New York: Tor, 1989), p. 52.

11. Ibid., pp. 370–371.

12. In this regard, I cannot resist passing along a perhaps apocryphal tale concerning the creator of Tarzan and John Carter. In Burroughs's sunset years, the woman who had for many years known herself to be the designated heir of his considerable estate made the mistake of referring to his fictions as escapist nonsense, supposing that he regarded them in the same light. He did not, and she was cut out of his will. Burroughs *believed* in his wish-fulfilling tales as fervently as Card believes in his. It is the secret of their success.

13. *The Encyclopedia of Science* (New York: St. Martin's, 1993).

14. Robert A. Heinlein, *Farnham's Freehold* (Riverdale, N.Y.: Baen, 1994), p. 296.

15. Hal Clement, *Mission of Gravity* (New York: Del Rey, 1984), p. 229.

Chapter 10

1. Louis Menand, "Hollywood Trap," *New York Review of Books*, September 19, 1996, p. 6.

2. Nick Lowe. "Mutant Popcorn," in *Interzone*, October 1996, p. 37.

3. Al Sarrantonio to the author, August 8, 1997. Quoted by permission.

4. Tracy Kidder, *The Soul of a New Machine* (New York: Avon, 1982), p. 242.

5. William Gibson, *Neuromancer* (New York: Ace, 1984), p. 63.

6. William Gibson, *Burning Chrome* (New York: Bantam, 1986), p. 178.

7. Mark Leyner, *My Cousin, My Gastroenterologist*, quoted from Larry McCaffery (ed.), *Storming the Reality Studio* (Durham: Duke University Press, 1991), p. 102.

8. Gibson, *Neuromancer*, p. 5.

9. Ibid., p. 25.

10. *The Encyclopedia of Science* (New York: St. Martin's, 1993).

11. Ray Bradbury. *Farenheit 451* (New York: Ballentine, 1953), p. 18.

ACKNOWLEDGEMENTS

My thanks are due, in the first instance, to Glen Hartley and Adam Bellow, my prime movers.

To Jim Allen, Al Sarrantonio, and a legion of other SF professionals, for their inside information—and with special warmth to Virginia Kidd, the mother of us all.

To my friends Charles Naylor and Jerrold Mundis, for services beyond the call of friendship.

And to my copy editor, Beverly Miller, and Carol de Onís of the Free Press, for their scruples and good judgement.

INDEX